One Cowboy's Dream

One Cowboy's Dream

John B. Kendrick

His Family, Home, and Ranching Empire

Cynde A. Georgen

THE
DONNING COMPANY
PUBLISHERS

Second Edition, Revised 2004

* * *

Published by
The Donning Company Publishers
184 Business Park Drive, Suite 206
Virginia Beach, VA 23462-6533

Steve Mull, General Manager
Barbara B. Buchanan, Office Manager
Richard A. Horwege, Senior Editor
Daniel Carr, Designer
Stephanie Danko, Imaging Artist
Mary Ellen Wheeler, Proofreader
Barbara Bolton, Project Director
Anne Cordray, Project Research Director
Scott Rule, Director of Marketing
Travis Gallup, Marketing Coordinator

Library of Congress Cataloging-in-Publication Data

Georgen, Cynde A.

One cowboy's dream : John B. Kendrick, his family, home, and ranching empire / by Cynde A. Georgen.—2nd ed., rev.

p. cm.

Includes bibliographical references (p.) and index.

ISBN 1-57864-239-6 (softcover : alk. paper)

1. Kendrick, John B. (John Benjamin), 1857–1933. 2. Legislators—United States—Biography. 3. United States. Congress. Senate—Biography. 4. Governors—Wyoming—Biography. 5. Wyoming—Politics and government. 6. Ranchers—Wyoming—Biography. 7. Kendrick family. 8. Ranch life—Wyoming—History. 9. Wyoming—Biography. I. Title.

E748.K37G46 2003
978.7'031'092—dc22
[B]

2003067435

Printed in the United States of America by Walsworth Publishing Company

Contents

Acknowledgments

In the eight years since *One Cowboy's Dream* was first published, we have not sat idle. New sources kept coming into our hands: a letter or two here, a diary or scrapbook there—all adding to the overall picture of John B. Kendrick, his family, his ranches, and his home. Sometimes we found information that validated what we already knew. Other times, we discovered facts that took our enquiries down entirely new paths, leading us toward new conclusions and new insights.

The author would like to thank the following organizations and individuals for their encouragement, assistance, and support. Without it, *One Cowboy's Dream* could not have been written in the first place, let alone revised:

- Past and present members of both the Trail End Guilds and the Sheridan County Historical Society.
- Former and current employees and interns of the Trail End State Historic Site: Nancy McClure, Dana Prater, Diane Marsden, Pamela Licht, Sarah Campbell Eaton, Irene Hause, Margot Liberty, Mary Butler-Atter, Hollie McHenry, and Sonny Reisch.
- The following members of the extended Kendrick family: the late Manville Kendrick, John B. and Cynthia Kendrick, Eula Harmon Hoff, the late Paul Hoff, Kendrick and Helen Harmon, Helen Kendrick Campbell, Chase and Diana Kendrick Untermeyer, John B. Kendrick III, Francis Kendrick, and Thomas White.

Cynde A. Georgen
Sheridan, Wyoming
2003

Part 1

Prelude to a Dream

No Lovelier Place on Earth

A BRIEF HISTORY OF SHERIDAN

In the last half of the nineteenth century, after the turmoil of the Civil War, millions of acres of land west of the Mississippi River were opened up to settlement. Homesteaders, many of them fleeing the overcrowded cities of the East, took advantage of those acres, plowing under some 225 million of them. Cattlemen also saw their opportunities spread out clearly in the vast open rangelands of Colorado, Wyoming, and Montana. By the 1880s, hundreds of "spreads"—both large and small—dotted the High Plains, providing livelihoods for wealthy cattle barons and simple cowhands alike.

In 1868, Wyoming Territory was created from the middle of this abundance. It soon became home to countless settlers and entrepreneurs looking to take advantage of everything the territory had to offer: mineral resources, natural beauty, and, perhaps the most important feature, wide-open spaces.

In the spring of 1882, Ohio native John D. Loucks came to northern Wyoming in his search for a place to make his home. His quest brought him into an area that only a few years earlier had been aflame with native unrest and political controversy. Though invited by the people of the tiny hamlet of Big Horn to settle in their community, he decided to head on to Montana. He only traveled a few miles, however, before he stopped, compelled to do

Sheridan County, 1911 (Trail End Collection)

so by the beauty of the land around him. In 1914, at the dedication of the monument marking the site of Sheridan's first cabin, Loucks described his early impressions of the area, before extensive settlement brought permanent changes:

> *I halted on what is now Court House Hill, tired and hungry. The sun was about an hour high; the evening air soft and pleasant, with an afterglow of golden hue; the hills were tinged with green. On a point across Little Goose was a small herd of buffalo, leisurely feeding; a little way up Big Goose was a small band of deer, roaming at will. To the west were the snowcapped Big Horn Mountains and the Big Goose Valley; to the north, Wolf Creek and Tongue River with their broad acres; to the east, the fertile valley of Prairie Dog, and to the south, the picturesque Little Goose.*

Walking down to the confluence of Big Goose and Little Goose Creeks, Loucks was struck with the certainty that this was the place he was looking for. That night, March 10, 1882, he drew the original plat of the town on a piece of brown wrapping paper. The next morning he contacted all the settlers he could find to advise

them of his plan to start a new town and name it after his Civil War commander, General Philip Sheridan. Loucks also contacted the land office in Cheyenne, which gave its approval to the name and locale. Nearly two years later, on March 5, 1884, Sheridan was incorporated by an act of the Wyoming Territorial Legislature. By the fall of that year, nearly all the lots on the original forty acres were occupied. Within a few years, the town's main street was lined with shops, and homes had sprung up all over the valley.

Main Street, Sheridan, 1888 (courtesy Wyoming Room, Sheridan County Fulmer Public Library)

Though Sheridan started out as a sleepy little community, the population increased dramatically after 1900—from 1,500 to 8,400—when underground coal mines were opened north of town. The mines lured hundreds of immigrant families to the county, all eager to take advantage of America's promised prosperity. Mostly from Italy, Scandinavia, Austro-Prussia, and a variety of eastern European states, the new settlers—both miners and railroaders—were very different from the English and Scots who had come in the 1880s to tend their herds on the plains of north-

Main Street, Sheridan, 1911 (courtesy Linda Prill)

central and eastern Wyoming. While the Italian coal miners and Japanese section gang workers, for example, lived in small, tight-knit camps reminiscent of their native villages, the cattlemen kept to spacious ranch houses built on their extensive holdings in the foothills of the Bighorn Mountains and on the plains of the Powder River Basin.

Another type of settler was the rural homesteader. Frequently occupied with the business of farming or small-scale ranching, homesteaders in the Sheridan area were luckier than those to the south near the town of Buffalo. Known as the Johnson County Cattle War, the difficulties in that area involved, in addition to problems with rustlers, the differences between homesteaders and cattlemen. The homesteaders wanted to farm their small parcels of land, fenced in and protected from intruders. The cattlemen, on the other hand, believed that the wide-open spaces of Wyoming should remain just that, wide and open, without any sign of fence or farmhouse. With increased crowding in the eastern cities and the rapid western expansion of "civilization," however, it was just a matter of time before the free-range cattlemen lost out.

Like John Loucks before him, John B. Kendrick first fell in love with the visual beauty of the Sheridan area. As Senator Joseph T. Robinson noted in his 1934 eulogy of Kendrick:

> *Passing over long trails, across plains inhabited only by wild Indian tribes, [Kendrick] sought an attractive place in which to make his home. When he beheld the tableland on which Sheridan is located and looked upon the distant Big Horn Mountains, he said: "There is no lovelier place on earth than this"* (United States Congress 1934, 40).

But Kendrick was not interested in beauty alone. When he first came to the area in the 1880s, he was impressed by Sheridan's perfect location, one ideal for commerce and industry. Though he had made his mark in cattle ranching, an endeavor in which settlement and homesteading were discouraged, Kendrick realized that it was the *people* coming into the Sheridan area that would make the difference between success and failure. By investing in real estate, banking, mining, and newspapers, Kendrick helped make Sheridan a prosperous community that worked with the cattle industry, not against it.

Even though he moved on to state and national prominence, Kendrick never forgot his adopted home, the little community where he established what he hoped would be his trail's end. In 1919 he noted in a letter to Sheridan Mayor H. A. Loucks:

> *I have always hoped that some time I could go back home and lend myself wholeheartedly to a scheme of making Sheridan*

one of the most beautiful and attractive residential towns, not only in the West, but on the face of the earth.

While many people may feel that Trail End contributes to the attractiveness of Sheridan, others over the years have been of the opinion that it was built to show off the Kendrick money. In a way, this is true. Starting out as he did, a penniless orphan with little education, it could be expected that John Kendrick would be proud of his accomplishments and want to show off a bit. To the Kendricks—John, his wife Eula and their children Rosa-Maye and Manville—Trail End was not just a home, but a symbol of the family's success. It was the result of hard work and enterprising spirit as well as a testimony to the extraordinary good fortune that aided John Kendrick in nearly everything he attempted.

Trail End (Harmon Collection)

The Logical Center of the Country

KENDRICK HISTORY IN CONTEXT

When he made the decision to build a town in northern Wyoming, John Loucks looked for a place where people would naturally want to settle. He found it along the banks of Goose Creek in what was then Johnson County:

> *All trails seemed to lead to this place. At a glance it was very apparent that this was the logical center of the country, all making a picture that, once impressed on the vision, could never be erased* (Loucks 1914).

John Kendrick saw Sheridan in the same way. By building where he did, he not only placed himself in the center of a vibrant local context, but in an important historical one as well. Most of the major events in northern Wyoming history happened within a few miles of Trail End.

South of the home, for instance, framed by giant silver birch trees, lies a sweeping panoramic view of the Bighorn Mountains. Most of the range, once known as the "Shining Mountains," is now part of the Bighorn National Forest and has long been of economic, cultural and recreational importance to the people of northern Wyoming and southern Montana. After more than twenty thousand years of Native American occupation, the Bighorns became known to Euro-American explorers and

Sheridan and the Bighorns, 1902 (courtesy Wyoming Room, Sheridan County Fulmer Public Library)

mountain men during the mid-to-late-1700s. It is not known which of these travelers first saw the range, but they were no doubt as impressed as first-time visitors are today by the snow-covered peaks soaring more than ten thousand feet above the valley floor. During the 1860s the foothills immediately east of these mountains saw violent conflicts between American Indians and the United States military as the Bozeman Trail brought miners and settlers through the last hunting grounds used by the Lakota, Cheyenne, and Arapaho.

Adjacent to Trail End, just south of the sunken rose garden, is Kendrick Park. Donated to the community by John Kendrick in 1905, it was known as Pioneer Park until renamed in Kendrick's honor in 1936. Above the park along Pioneer Road is a pasture stocked with elk and buffalo, another Kendrick gift to Sheridan.

Across the valley, also looking south, are the Sheridan Munici-

PK Ranch rodeo, 1929 (Manville Kendrick Collection)

pal (Mount Hope) Cemetery and the Sheridan County Airport. Beyond these, about seven miles south, is the community of Big Horn. For several years during the 1880s, Big Horn rivaled Sheridan as the population center of the county. When the railroad bypassed Big Horn in favor of Sheridan, however, Big Horn quickly lost most of its inhabitants and businesses. By the 1920s, it served primarily as a service community for area ranches and as the home base for a number of successful polo teams. Several of Big Horn's ranches, including those of the Wallop and Moncreiffe families, were founded by younger sons of British nobility who came west in the 1880s to seek their fortunes. The British connection is still evident with Great Britain's Queen Elizabeth and Prince Philip among those who have visited in recent years.

West of Sheridan lay the fertile lands of the Big Goose Valley. In the late 1800s and early 1900s, this valley was full of small farms producing everything from sugar beets and wheat to apples and

Kendrick cowboy in Powder River country, 1916 (John B. Kendrick II Collection)

watermelons. Spectacular cattle ranches lay along the base of the Bighorns, including Wyoming's first dude ranch, Eaton's, located in Wolf, Wyoming.

North of Sheridan, past Trail End's towering blue spruce, are the communities of Parkman, Ranchester, and Dayton. In the 1890s, the longest tie flume in the United States ran into Dayton from the Bighorn Mountain logging camp of Woodrock. The ties were used to build tracks for the Burlington Railroad as it passed through Sheridan County on its way to Montana.

Also to the north are the remains of several mining towns including Monarch, Acme, and Dietz. Flourishing between 1900 and 1930, these communities have long since disappeared. Average

travelers along Interstate 90 have no idea that millions of tons of coal were mined out from under the hills surrounding them, providing jobs and steady incomes for thousands of immigrant miners, as well as the entrepreneurs, including John Kendrick, who owned the mines.

Trail End's front (east) porch overlooks the confluence of Big and Little Goose Creeks, the original site of the town of Sheridan. In June of 1876, six years before Loucks' arrival, the military command of General George Crook departed northward from this fork to engage the Lakota and Cheyenne at the Rosebud Battle. On June 17, Crook's forces were turned back by the same warriors who would defeat George Custer and his troops at the Battle of the Little Big Horn one week later.

Beyond the high hills east of Sheridan lies Powder River country. Vast herds of Texas longhorn cattle followed the banks of Powder River on trails leading to the grasslands of Wyoming and Montana. John Kendrick ranched in this area, utilizing the bountiful shortgrass rangelands adjacent to the waters of Powder River to create a ranching empire of just under 210,000 acres.

Part 2

The Dream Weavers

The Call of the Range

JOHN BENJAMIN KENDRICK

The youngest son of a pioneer Texas family, John Benjamin Kendrick was born on September 6, 1857, in Cherokee County, Texas. His father, John Harvey Kendrick, had arrived in Texas in the early 1830s as part of a large extended family. They founded the town of Griffin, named after their former place of residence in north-central Georgia. Once established in Texas they were involved in a variety of agriculture-based enterprises, cotton production being primary among them.

John B. Kendrick, 1915 (Manville Kendrick Collection)

John Kendrick's mother Anna Maye was a native of County Sligo, Ireland. The daughter of a wheelwright, Anna came to America to live with her married sister in New Orleans. According to Kendrick family friend J. Vinson Wiggins:

> *An aunt of Governor Kendrick's visited in New Orleans and got acquainted with this estimable family, and taking a fancy to this beautiful Irish girl, later had her to visit them on their plan-*

tation in Texas. That visit resulted in her marriage to Governor Kendrick's father ... who had a year earlier lost a former wife, and by whom he had several children.

John Harvey's first wife, Hettie Humphries Smith, had died in childbirth in 1855, leaving five children ranging in age from newborn to ten years. Believing he'd found a good mother for his children, Kendrick married twenty-three-year-old Anna Maye in 1856. Fourteen months later they welcomed their first child, John Benjamin. Two years later, daughter Rosa was born.

John B. Kendrick, about 1870 (Manville Kendrick Collection)

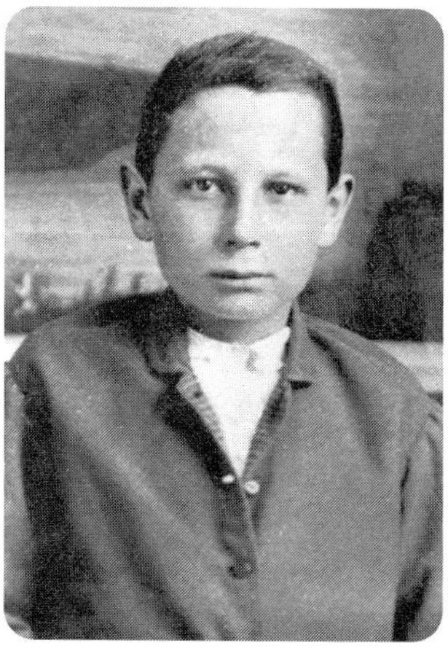

Rosa Kendrick, about 1870 (Manville Kendrick Collection)

The Kendricks did well in Texas for many years. After the ravages of the Civil War and their unsuccessful backing of the Confederate cause, however, any wealth the Kendrick family might have had was gone, and so were both of young John's parents. John Harvey Kendrick had drowned in 1860 while trying to ford a river near his home in Griffin. After his death, Anna took care of her own children as well as her stepchildren. Three years later, at the age of twenty-nine, Anna herself was dead. John was six, his sister only four.

After they were orphaned, John and Rosa stayed first with a childless aunt and uncle, and later with half-sister Mary Kendrick Reavis. John lived with this family for eleven years and later commented that Mary's husband, Thomas Harrell Reavis, had a "profoundly positive" influence on his life, instilling values of thrift, abstinence, and hard work. Shuttled back and forth as they were from one relative to another, the Kendrick children developed resiliency and patience, qualities that would serve them well in the future.

Kendrick rarely spoke of his early childhood, preferring to begin his personal anecdotes around the time he first went to work as a fledgling cowboy. One of his earliest jobs, around the age of fifteen, was breaking horses for room and board. In 1879, at the age of twenty-two, Kendrick hired on with the Snyder-Wulfjen Brothers of Round Rock, Texas, to help move a herd of three thousand cattle from Matagorda Bay (on the Gulf of Mexico) to the grasslands of Wyoming. As an inexperienced hand, he made only thirty dollars a month. During that trip, Kendrick fell seriously ill and nearly died:

John B. Kendrick, 1877 (Manville Kendrick Collection)

> *... the foreman of the gang of cowboys came over to the wagon and said, "You'd better let us leave you here to get well. The doctor says you won't live if you go on." And I said, "I know, the boys talked about it to me yesterday. But I've made up my mind to go on. I think I'll get well, but if I don't it won't be much trouble for you to bury me along the trail." So they let me go on with them across the Red River. In front of us were hundreds of uninhabited miles to Wyoming, where the cattle were to be driven. As we rode up the long slope out of the river valley ... I propped myself up on my elbows with an effort and took one last look back across the river upon Texas* (Gilbert 1928, 14).

Kendrick gradually recovered and completed the long journey to Wyoming, traveling through Oklahoma Territory, Kansas, and Nebraska. Along the way he experienced the worst of what the new land had to offer: drought, flood, storm, and stampede:

> *Anything at night almost, a stumbling horse, the odor of some wild animal or a blinding flash of lightning would start the*

herd to running. Then the cowboys would have to follow. Sometimes the cattle would run until morning. Or they might stop now and then for a few minutes. Whereupon, of course, we would halt with them, the rain meanwhile coming down in torrents and the night being so black that nothing could be seen but the electricity on our horses' ear or the lightning wriggling over the ground like illuminated serpents (Morrow 1917).

Like many of the cowboys with whom he rode, Kendrick's formal education was incomplete; he did not continue regularly after the fifth grade. Unlike the others, however, he more than made up for his lost lessons on the trail and during roundups. As cowboy A. G. Crane noted:

It was the custom in those days for the cowboys to proceed to shoot up the town, to gamble, and consume all the liquor they could after the cattle were loaded and shipped. It sometimes took two or three days for the boys to sober up. While the cow-punchers were off on their spree, and after Kendrick had transacted his business, he would go to the roundup wagon and put in his spare time reading and studying from books (always good literature), that he carried in his saddle pockets (Deming 1932).

Kendrick later estimated that he only spent about fifty dollars a year on himself, and most of that for books. With his remaining wages he bought a few head of cattle which his employer, Charles Wulfjen, allowed him to graze with the main herd. By 1882, when Wulfjen's holdings were absorbed by the massive Converse Cattle Company, Kendrick was able to sell his own cattle to that company, using the profits to begin another herd. He managed his

John B. Kendrick (center), about 1883 (courtesy Anna Miller Museum, Newcastle, Wyoming)

herds and finances so successfully that when the disastrous winter of 1886–87 brought an end to the golden age of the cattle baron, Kendrick was in the financial position to take advantage of what was for nearly everyone else an impossible situation. In 1887 he signed on as superintendent of the Converse Cattle Company. Two years later he moved the entire operation from eastern Wyoming to the banks of Hanging Woman Creek in south-central Montana.

John B. Kendrick holding infant son Manville and daughter Rosa-Maye at OW Ranch, 1900 (Hoff Collection)

By 1900, he was not only owner of the Converse outfit, but a successful businessman as well.

In 1889, speculating on the coming of the railroad to Sheridan, Kendrick formed a partnership with Nebraska banker A. S. Burrows to start the town's second bank. According to the *Sheridan Post*, Burrows provided the management skills while Kendrick contributed a good deal of the start-up money as well as considerable acres of investment land:

> *It will be seen at once that [Mr. Burrows] is a business man of no ordinary capacity, and that he will be a valuable acquisition to our business circles. Associated with him as president of the bank, is Mr. John Kendrick, the efficient general manager of the Converse Cattle Company … [Kendrick] has resided in Wyoming for the past eleven years, and is a large holder of Sheridan real estate.… Together they will make a strong firm, and we bespeak for them a liberal share of the patronage of our people.*

The partnership dealt primarily with mortgage loans on farmland and Sheridan real estate. Kendrick's line of reasoning was that when the railroad came to town (no "ifs" allowed), prices would skyrocket and huge profits would be made. His

gamble paid off after the Burlington Railroad's first train arrived in Sheridan in 1892. When new settlers started coming to town, Kendrick and Burrows sold their lots for many times the original purchase price.

A. S. Burrows and John B. Kendrick at OW Ranch, about 1920 (Hoff Collection)

After a few years spent surviving depressions, recessions, and inflations—and after serving a brief term as president of the First National Bank of Sheridan (between 1900 and 1902)—Kendrick liquidated his banking interests and went back to expanding his cattle operations, based at the OW Ranch in Big Horn County, Montana. He also began looking toward his future.

Since its earliest territorial days, Wyoming's political positions had been manned and managed by the Wyoming Stock Growers Association. If someone wished to run for office, he had to have the endorsement of the cattlemen. It was even better if the candidate was himself a member of the organization.

John B. Kendrick, OW Ranch, 1895 (Manville Kendrick Collection)

Being primarily ranch owners rather than ranch hands, WSGA members traditionally frowned on cowboys owning their own herds (thus making them eligible for membership). According to the organization's records, Kendrick was the first "cowboy" granted membership in the WSGA and the only one exempted from the Association's proscription against cowboys owning livestock (*Sheridan Press* 1933). Since one of his first jobs in Wyoming was deemed managerial, as foreman of the 1884 Wulfjen cattle drive, Kendrick was apparently not considered to be a common cowhand. It also did not hurt that his application for WSGA membership was endorsed by future father-in-law Charles Wulfjen, himself a charter member of the group.

Although a member since the 1880s, Kendrick did not attend a state WSGA convention until 1910. He went on to serve as president of the group in 1912–1913 (a post held by his son nearly forty years later). He also served a term as president of the American National Live Stock Association from 1919 to 1921.

Kendrick's reputation grew large in cattlemen circles, but he remained a political unknown until 1910 when he was elected, with the support of the WSGA, to the Wyoming State Senate by the voters of Sheridan County (this despite the fact that his Sheridan home was not yet complete and he spent the majority of his time on his Montana ranch). Three years later he ran for the United States Senate seat held by Republican Francis E. Warren. Although he lost the election, Kendrick gained much in the way of public recognition. By 1914 he was able to win the governorship, only the second Democrat in Wyoming history to do so.

State Senator John B. Kendrick at work in Cheyenne, 1910 (Harmon Collection)

Although his actions clearly indicated that he was willing, Kendrick wanted his constituents to know that he was making a tremendous personal sacrifice by agreeing to a life of public service. In his letter of acceptance delivered to the Democratic State Committee in 1914 following his nomination for governor, Kendrick voiced his "reluctance" to step into the political spotlight:

> *Very naturally, I am disinclined to become a candidate for public office, and but for the obligations of citizenship in-*

Governor John B. Kendrick delivering inaugural speech, Cheyenne, 1915 (Schimitz, Cheyenne; Hoff Collection)

volved, would much prefer the pursuits and privileges of private life. Under our form of government, [however,] I am unable to reconcile my ideals of good citizenship with a refusal to accept office.

Despite being a Democrat in a traditionally Republican state, Governor Kendrick worked harmoniously with the overwhelmingly Republican legislature on a variety of reform-minded bills. During his brief tenure in office, Kendrick signed legislation providing for a workers' compensation law, a mothers' pension act and the establishment of a state public utilities commission.

Kendrick spent only two years as governor of Wyoming before state and national Democratic leaders called on him to serve the party in a higher capacity. In October of 1916, *The Commoner*, a national magazine published by perennial presidential candidate William Jennings Bryan, outlined the reasons why:

> *Governor Kendrick … is a self-made man, educated and polished. His people believe in him, they admire him, and without regard to political preference, will vote for him. His long, active residence has enabled him to learn the needs of the state at first hand, and he has proven that he is an able advocate in helping to supply those needs. Wyoming will have a strong, forceful man in the United States senate, if she sends Governor Kendrick to Washington, and the indications are better every day that she will do this very thing.*

In 1916, while still governor, Kendrick became the first United States Senator from Wyoming to be elected by popular vote—up until this time, the individual state senates elected United States senators; the responsibility was passed to the general electorate in 1913 with passage of the Seventeenth Amendment to the Constitution. Nominated in a write-in campaign, Kendrick defeated Clarence D. Clarke, a conservative Republican who had served in the Senate for nearly twenty-two years. Honor though it was to be elected to the Senate, Kendrick later said that he would have liked to have kept his old job. As he told Kansas schoolgirl Helen Robinson in 1932:

> *In 1916, when I had served only half my term the people of the State requested that I become a candidate for the United States Senate, but I declined because I preferred to continue to serve out my term as Governor. My name was written in the ballots and I was elected to the Senate and resigned as Governor to*

Governor John B. Kendrick with staff, 1917 (Schimitz, Cheyenne; Trail End Collection)

assume my duties in Washington in 1917. From this you will see that I "became interested in the Senate" rather involuntarily and was really drafted into service.

John Kendrick's first official duty as a newly elected United States senator was to sign the Declaration of War against Germany in 1917. This was one of his most difficult duties, as he had a healthy young son who, if the war lasted long enough, might someday be old enough to be drafted.

Like other politicians concerned about the troops, Kendrick braved U-boat infested waters to sail overseas on a fact-finding mission to the front lines. His reports, along with those of other official visitors, helped Congress make informed decisions on funding and staffing needs. After experiencing a submarine attack, an air raid in London, a formal dinner with General John J. Pershing, and an inspection tour of the military zone, Kendrick visited some of the soldiers near the front lines. As he wrote in an unidentified newspaper in 1917, he was profoundly moved by what he saw:

American Field Artillery in the Argonne, 1918 (Dab Maury; Manville Kendrick Collection)

> *We traveled over hundreds of miles of the front and had a glimpse here and there of the actual line of battle … It goes without saying that no man who has ever really glimpsed the war in Europe can be quite the same again. The effect of it is one well calculated to sober the mind of most anybody.*

Upon his return, Kendrick had words of advice for the folks back home concerning how they should communicate with their loved ones overseas. He told his niece, Josephine Kendrick Woodrel, that letters to her brother serving at the front should not be melancholy, but upbeat and supportive:

> *Instead of writing doleful and pessimistic letters, do not fail to remember to have your letters breathe words of good cheer and encouragement.… While telling him how much you miss*

him and how glad you will be to have him at home, do not overlook the very important responsibility of expressing your gratitude and appreciation at the fact of having a man in the family who was man enough to fight for the principles and for the ideals of his country.

Kendrick himself was widely regarded as a man willing to fight for his principles and, as a conscientious member of Congress, did so by taking part in debates on many different topics. He did, however, have one primary area of interest: the irrigation and reclamation of arid lands. In his relentless pursuit of water for thirsty crops and herds, Kendrick-backed legislation resulted in appropriations for Wyoming's Alcova Dam (near Casper) and Arizona's Boulder Dam (now called the Hoover Dam).

Although the greening of the West was his primary focus, Kendrick was not blind to the effects such legislation could have on the environment. He fought hard, for example, to preserve the sanctity of at least one water source when it was threatened by irrigation legislation:

> *Kendrick has stood like a rock against [Montana Senator Thomas Walsh's] proposal to authorize the damming of Yellowstone Lake for irrigation purposes [having] seen what a desolation the dam at the source of Snake River made of Jackson Lake—water-killed forests, dead and ghastly, and a fine sandy shore replaced with mud and marsh and swamp—and he was determined that nothing like that should ever come to the Yellowstone* (Knappen 1924, 51).

Though he had a number of scandals in his own political career, Kendrick was known as a reformer intent on rooting out

Senator John B. Kendrick (lower right) and colleagues, about 1930 (News Picture Service; Manville Kendrick Collection)

corruption. As soon as he entered the Senate in 1917, for instance, Kendrick became involved in President Woodrow Wilson's crusade for regulation of the meatpacking industry. In 1922, he was instrumental in starting the inquiry that eventually led to the uncovering of the infamous Teapot Dome Scandal, an administrative disaster for President Warren G. Harding involving bribery and fraud. Although others generally receive the credit for exposing and investigating the debacle, it was Kendrick who called the attention of his senatorial colleagues to the possibility of illegal actions revolving around Wyoming's Naval oil reserves. Senator Robert LaFollette (R-Wisconsin), the man generally given the credit for exposing the scandal, recognized this contribution:

> *The able Senator from Wyoming took the initiative in this matter and by his resolution focused upon it the attention of the public. For that action he deserves the gratitude of the country.... I propose to support and sustain him in the stand he has taken* (*Congressional Record* 1922).

Such sentiments were probably very much appreciated by the junior Senator from Wyoming because, like many a whistleblower, Kendrick was not treated so kindly in his own home territory. In the tradition of William Randolph Hearst and the large Eastern newspapers, the region's daily and weekly papers took strong positions on public issues, with editorial opinions sharply divided along political party lines. Harking back to rumors circulating during his gubernatorial days, for instance, one Colorado newspaper wrote:

> *Hon. John B. Kendrick of Wyoming is given much credit for "exposing" the Teapot Dome lease. Anyone who is conversant with the past record of that individual should know without a second thought that the charges were pure "bunk." If ever a state or nation was robbed of its oil interests, John B., et al., while Governor of Wyoming, sure cleaned "the kettle." It isn't possible that John has changed and probably the only thing hurting him is the fact that he overlooked "taking" the Teapot "in" while still in position to do so. He probably considers it a blot on his fair name, as he certainly got everything else he could lay his hands on. He should be the last to cry, "holier than thou"* (*Golden Republican* 1922).

The "past record" referred to by the *Golden Republican* involved Kendrick's land acquisition practices. He was accused at various times of having illegal fences around government land, obtaining state lands while serving as governor, and using employees and "dummies" to obtain homestead lands from the federal government. Although he had plausible explanations and excuses, there is little doubt Kendrick was guilty of committing the actions of which he was accused. But he was apparently not acting alone.

As Wyoming historian T. A. Larson noted, "It may be supposed that he [Kendrick] was like other big cattlemen of his day with respect to these practices" (Larson 1965, 328). It is probably for this reason that the allegations against him did not become serious political stumbling blocks. Kendrick's opponent in the 1912 election, for example, rancher-politician Francis E. Warren, was himself accused of illegal fencing, perjuring himself to obtain a land title, and improperly benefiting from the lease of state lands. Unless they wanted to draw more attention to the charges against their favorite son, the Republicans could not press the same issues too strongly against the Democratic Party's candidate. As for the voters, they ignored the whole issue, voting for both Kendrick and Warren in numerous elections.

Throughout his career, Kendrick frequently played down his own importance and abilities. In a 1929 interview with Fraser Edwards of the *Washington Herald*, he noted:

> *I've never qualified as an orator, but I've gotten so that I can speak anywhere on any subject except in the United States Senate—and nobody there ever listens to what you have to say.*

This was not altogether true. Reporters were always listening, and for most of his political life, Kendrick had problems with the press. Scarcely a Wyoming newspaper was published that did not include a story, quote, or cartoon about him on an almost daily basis. He was constantly accused, defended, praised, and lampooned by both large and small publications. In 1922, when

Kendrick found himself being supported by a coalition of left-wing activists, an amused editor at the *Douglas Enterprise* noted:

> *It is extremely mirth-provoking to think of the resounding, bombastic oratory and silly chatter of the would-be saviors of the country, who lambasted all corporations and rich men (excuse us, republicans only) at the Democratic-Progressive-Non Partisan-I.W.W. convention which met here last week and then have them wind up by endorsing Millionaire John Kendrick, the wealthiest man in the state....*

Early in his political career, in an effort to protect his interests and those of the Democratic Party, Kendrick bought controlling interest in the *Sheridan Enterprise* and the *Cheyenne State Leader.* Such connections allowed him and his party to balance the harsh coverage of the state's leading Republican papers, including the *Sheridan Post* and the *Cheyenne Tribune.*

Despite such problems, Kendrick was reelected for two more terms in the Senate, handily defeating his Republican opponents in 1922 and 1928. It was suggested by some in 1933 that Kendrick be nominated by both the Republican and Democratic Parties and then unanimously acclaimed senator in 1934. No wonder: Kendrick was influential in Washington and few Republicans felt up to the task of challenging him. Although he had already announced his decision not to seek reelection in 1934, few believed him. Why, they wondered, would Kendrick not wish to continue in office? He was, after all, the chairman of the committee on agriculture, the second-ranking member of the Appropriations Committee, and was about to preside over the Committee on Public Lands. In addition, the vital seventy-six-year-old was considered by many Washington politicos to be a prime Democratic

Political cartoon, Casper Tribune, *1922 (Trail End Collection)*

candidate for vice president in the 1936 election. It appeared that more than a simple retirement awaited John Benjamin Kendrick.

Before such a path could be explored, however, Senator Kendrick's life was over. By the fall of 1933, those around him had noted that the Senator looked exhausted; he was encouraged to take some time off and relax. Unfortunately, he did not heed the advice. Kendrick was rushed to Sheridan County Memorial Hospital the first night of November after becoming ill and complaining of a headache while working at his downtown office. He slipped into a coma and died two days later, November 3, 1933, surrounded by his immediate family. A cerebral hemorrhage was listed as the cause of death.

Following a funeral attended by politicians and cowboys from around the state, Kendrick was buried in the Sheridan Municipal Cemetery. An editorial tribute in the *Sheridan Morning News* (one of hundreds appearing in local, regional and national papers) noted:

> *Born of the West, the Kendrick name is enshrined in Wyoming as symbolical of a human effort, unceasing, untiring, sincere in behalf of the people whom he represented.*

Kendrick's contributions to his community, state and nation went beyond the political. As part of his desire to beautify his adopted city, for example, he became involved in the establishment of city parks and recreation areas throughout Sheridan County. As he told H. A. Loucks in 1919:

> *... you are entirely right in your impression that I am interested in seeing every possible step taken that will increase the beauty of Sheridan and bring comfort and happiness to its citizens. I know of nothing that offers a more direct opportunity for sane and healthy recreation and comfort than such playgrounds located right in the town where people can get to them without any delay or any long journey.*

A man of action, Kendrick's deeds supported his words. In 1905 he donated seventeen acres adjacent to Trail End for use as a municipal park. In 1930, he gave over five hundred acres west of Sheridan exclusively for use as a city golf course. He even spoke of connecting Sheridan's various recreation areas by building a green parkway extending along both sides of Big Goose and Little Goose Creeks.

Kendrick was also interested in the growth of Sheridan, not only as a supply stop for area farmers and ranchers, but as a regional center for business and industry. Toward that end he served on the boards of directors for several industrial concerns, including Sheridan Flouring Mills and Sheridan Iron Works. He also helped develop downtown Sheridan by financing the construction of several imposing retail and office structures. Among these are the Sheridan Commercial building at 303 Broadway (rebuilt by Kendrick following a disastrous fire in 1916) and several Main Street structures. All can be recognized by the distinctive "K" over the front door.

Regardless of whatever else he did with his life, John Kendrick was first and foremost a cattleman. He traveled back to the OW

Sheridan Commercial, built by John Kendrick, 1916 (Mills Company; courtesy Linda Prill)

John B. Kendrick (left) in OW Ranch corral, about 1906 (Hoff Collection)

Ranch whenever his busy Washington schedule would allow. He championed the political causes of the cattlemen and wrote extensively of their problems, often using anecdotes from his own days in the saddle to make his point.

In 1932, Kendrick received a letter from Tom Bell, an old cowboy acquaintance. This was not unusual, as many of his former coworkers wrote after seeing John's name in the papers. As always, the Senator drafted a detailed personal response, keeping a copy for his files. In his reply, Kendrick recalled their early days together and reflected on the turns his life had taken:

> *While we have an unusually fine plant, including the range and the ranches, if it were to do over again I should hesitate a long time before I would buy lands and build ranches on*

which to handle cattle on an extensive scale. There is too much expense and too much grief, though in all candor this seems to be the exact experience of almost every other line of business.

Though Kendrick may have occasionally voiced regrets about his chosen path, such complaints were few and far between. As he said in a 1924 article written for *The Country Gentleman*:

But notwithstanding all its drawbacks, the cattle business is the most deeply satisfying business in which men engage. No other business so fully occupies a strong man's full powers of body and mind. Cattle ranching still possesses great allurements and some of its traditional romance.... Those who have grown up with the business can never escape the call of the range.

It was that "call of the range" that drew John Benjamin Kendrick up from the shores of Matagorda Bay, north over the dusty trails of Oklahoma and Kansas and into the knee-high grasslands of eastern Wyoming. It was the call of the range that moved him even farther north to Hanging Woman Creek, a place where cattle, children, and aspirations could be raised up healthy and strong. And it was the same call that beckoned him into the world of politics and away from the rangeland he loved so much. Kendrick knew and accepted the reality that, in order to protect the range, he inevitably had to leave it.

Senator John B. Kendrick, 1932 (Harris and Ewing, Washington, D.C.; Manville Kendrick Collection)

Hearty in Nature and Heart

EULA WULFJEN KENDRICK

It is said that Western men admire women with spunk. If this is true, Eula Wulfjen Kendrick must have had many admirers among the cowboys and politicians with whom she was associated, for spunky more than describes this hearty Western woman who took what life had to offer and nearly always changed it into what she desired it to be.

Born in the Texas ranching community of Round Rock on April 26, 1872, Eula was just fifteen months old when she moved with her family to Cheyenne, the bustling capital of Wyoming Territory. Her father, Charles William Wulfjen, was then in the process of becoming one of the largest cattle ranchers in the region. The family moved to Austin, Texas, in 1882, but headed north again in 1884 to settle permanently in Greeley, Colorado.

Eula Wulfjen Kendrick, 1916 (Brown, Denver; Trail End Collection)

No matter where they lived the Wulfjens maintained a genteel lifestyle. A Civil War veteran who had fought for the Confederacy, Charles Wulfjen had a reputation for honesty and fair dealing. His growing importance in the business circles of Cheyenne and Greeley afforded his wife and family a certain measure of social prominence:

The growing beauty of our town is yearly attracting more and more people of wealth and refinement to settle here, even when their business is somewhat remote. This may be said to be the case with C. W. Wulfjen, whose principal headquarters for his large herd of cattle are in Texas. His daughters, talented in both music and elocution, are quite an accession to the cultivated society of the place (Boyd 1890, 414).

Eula Wulfjen, about 1884 (Kirkland Studio, Greeley; Manville Kendrick Collection)

Eula and her older sister Mattie attended finishing school in Boulder, Colorado, where they rounded out their education with the study of music, painting, and elocution. (Mattie so excelled at her studies of language and vocal production that in the 1890s, prior to her marriage to engineer Francis Williams, she toured as an elocutionist with a Utah-based theater company.) Clarence Wulfjen, Eula's only brother, also received a fine education in preparation for the day when he would take over his father's ranching business. Two other youngsters, Edna and Hazel, died of typhoid before they could enjoy the fruits of their father's efforts.

Below: Charles W. and Ida Peeler Wulfjen, about 1869 (Hoff Collection)

Eula's mother, Ida Josephine Peeler Wulfjen, was born in Florida, the daughter of an itinerant preacher and Confederate Army chaplain. She was not a strong woman and her two surviving daughters were raised to take responsibility for many of the household chores and duties usually performed by the woman of the house. The girls took care of featherbeds, washed dishes, dusted furniture, and kept their clothes in order. Eula later recalled:

> *When I was a mere child, my thrifty mother instilled into me habits of industry by assigning little duties and teaching the proper way to do them. I was kept busy, a habit which I have always followed* (Painter 1928, 201).

Although the girls were instructed in general housekeeping, there was nearly always some sort of hired help in the kitchen. When there was not, Charles Wulfjen did the cooking himself. The girls spent little time studying the culinary arts, an oversight Eula would later regret.

Like many ranching families even today, the Wulfjens lived in town during the winter and on the ranches in the summer. Being active children brought up to appreciate the benefits of fresh air and exercise, Eula, Mattie, and Clarence looked forward to the end of school and the start of summer, even if there were certain dangers involved:

> *During the winter they led a humdrum existence; the fact that Ida's pet deer ate the fringe off the baby carriage, is recalled as an exciting episode! But in the summer, when they always moved out of town, life took on a more exciting tinge. Eula's own first definite memory is of Mule Shoe Ranch on Sabille Creek [in southeastern Wyoming], where the Indians, who were still numerous and uncowed, swooped down in the middle of the morning one day when her father was absent, and stole the relay horses that he had left in the corral* (Keyes 1931, 40).

In later years, the family spent summers at the ULA Ranch, named for Eula, near Lance Creek in eastern Wyoming. There, when she was almost seventeen, Eula was reintroduced to one of her father's former employees, a cowboy named John B. Kendrick. She remembered meeting him before: at age seven she had climbed into the lanky cowboy's lap and announced that when she was old enough she intended to marry him (Keyes 1931).

Eula Wulfjen Kendrick wearing wedding dress, 1891 (Fredrick's Portrait Gallery, New York; Hoff Collection)

She must have been one determined little girl because, at the age of eighteen, Eula married her thirty-four-year-old cowboy. The wedding, described by the *Greeley Tribune* as "one of the happiest events in Greeley's history," took place on a Tuesday afternoon in January 1891. Eula was attended by her sister and best friend, while John was accompanied by Addison Spaugh and George Bissell (described by reporters as "two prominent cattlemen of Wyoming"). The wedding was an elaborate affair, one which caused the newspaper reporter to gush:

> *... then the bride and groom [entered], making a perfect picture of beauty, all standing under an arch of evergreens and flowers illuminated with colored electric lights. In the center hung a large bell of evergreens and flowers with an electric light suspended from the center. The altar was a scene of artistically arranged flowers, paintings, and banners, exquisite taste being displayed by loving hands in the whole arrangement. The entire ceremony was perfectly beautiful. The bride wore a costly and beautiful traveling costume. The groom wore the conventional black.*

Following a brief reception at the Wulfjen residence, the newlyweds left for New York on the afternoon train. With the financial resources gained through a lifetime of frugality, John was able to make the honeymoon one of comfort and enjoyment for his new bride. She later noted in her memoirs:

We stopped 1st at Brown Palace Hotel, Denver, Colorado, then Paxton Hotel (the old one of that name), Omaha, Neb., then to the famous old Cattleman's hotel in Chicago, the Palmer House (bought furniture for ranch home in Chicago), then to Niagara Falls, then to Albany, N.Y., and down to New York City by boat, then to Philadelphia, then to Washington, D.C., and back to Greeley.

While in Washington, the couple had a glimpse of their future life when they were admitted to the Reserved Gallery of the United States Senate Chamber at the invitation of Wyoming Senator Joseph M. Carey.

Having never been east of Kansas City, the two-month trip was an eye-opening experience for Eula, whetting her appetite for the finer things in life. When the trip was over, however, Eula was brought back down to earth by the reality of her new home: a mud-chinked log cabin fifty miles from the nearest town.

It would actually be several months before Eula would get to live in that cabin. Upon their return, Eula stayed at her parents' home in Greeley while John went to Montana to finish construction. He felt that the rough bachelor digs he had left behind were not good enough for his cultured bride. It was a lonely time for both John and Eula and letters flew back and forth between them. For a man accustomed to solitude, separation from a loved one

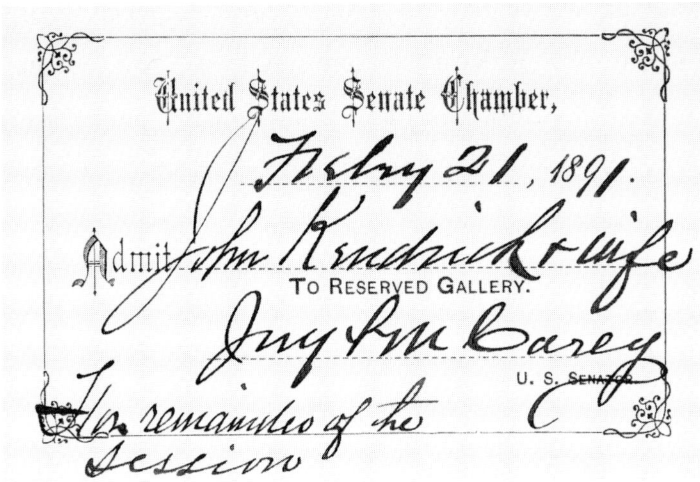

United States Senate Chamber,
Feby 21, 1891.
Admit John Kendrick & wife
To Reserved Gallery.
Jny M Carey
U. S. Senator
For remainder of the Session

Gallery Pass, U.S. Senate Chamber, signed by Senator Joseph M. Carey of Wyoming, 1891 (Manville Kendrick Collection)

was a new thing for John. He expressed his loneliness eloquently and often during this period. On March 21 he wrote:

> *Do you miss your old man? Not one half so much as I miss "the girl I left behind me." Somehow the feeling of loneliness is inexplainable. Everything lacks interest: the scenes along the road, the different views of the snow peaks of the Big Horns, things that I used to enjoy so much.*

Carpenters were hard to come by, and John found that finishing a cabin alone was not only hard, but took valuable time away from his work with the herds. By the middle of April, his frustration was beginning to come to a head:

> *Although I have worked almost day & night through rain and sunshine since my return, preparations for your coming progress very slowly.... If I thought you would be contented and happy with me here I would go down to Greeley and carry you up myself rather than leave you there, house or no house.*

By the first of May, the cabin was still not finished, but their forced separation was. Fed up with living over four hundred miles apart, Eula announced to her husband that she *was* going to Montana, even if she had to sleep on the floor and cook for herself. This response delighted John to no end:

> *You can never know how many false notions you have driven from my mind in your proposal to come out and do your own cooking—not that I want you to do it—but I did want so much for you to show the spirit of a true little wife and helpmate and the one thing needed to fill my cup of happiness you have supplied.*

Main house, OW Ranch, Montana, 1916 (John B. Kendrick II Collection)

The OW Ranch in southeast Montana was Eula's home for the next eighteen years, during which time she did indeed act as the helpmate John required. Though far from friends, she had no time to be bored: she cooked, cleaned, ironed, sewed, and helped

John with his ever-growing livestock interests. According to author Frances Parkinson Keyes, a close friend, Eula looked upon this time as "a period of development":

> *... the hard work, the primitive equipment, the extreme isolation, did not daunt the high spirits of the bride in the least. ... She had taken a short business course one summer in Greeley; and now, seeing what an effort it was for John Kendrick to do the bookkeeping, when he came in exhausted after a hard day in the open, she offered to keep the ranch accounts. Since they were not yet "on their own," and he felt acutely his personal responsibility to his employer, he hesitated to let her do this; but they compromised by deciding that she should balance them* (Keyes 1931).

Eula Wulfjen Kendrick with Manville and Rosa-Maye, OW Ranch, about 1905 (Hoff Collection)

Eula had the personal philosophy that whatever one disliked doing, one should learn to do well. Having never learned to cook as a young girl, she made herself study cooking when it became apparent to her that good food was required to keep skilled workers on the ranch. Occasionally there would be a hired cook, but more often than not the burden fell to Eula. Although she never enjoyed it, Eula learned to prepare meals and once cooked for eight men on a cattle drive rather than stay at home alone.

After her children were born, Eula took on the responsibility of providing them with an education. Without assistance or formal teacher training, she prepared Rosa-Maye for the seventh grade and Manville for fifth. As Eula noted in her journals, the two children learned many lessons beyond reading, writing, and arithmetic:

Rosa-Maye and Manville Kendrick, OW Ranch, about 1906 (Manville Kendrick Collection)

> *There were no near neighbors or schoolhouse nearby, hence no playmates, so mother had to be teacher and companion.... As the afternoons were given over to long horse back rides, the lessons were completed by noon, and then the fun began. If it was summer, the three would go for a swim in a not-too-deep hole in Hanging Woman, the little creek that wiggled along circuitously from the divide. In the early evening, too, they often went shooting for prairie dogs—surprising one (as well as themselves) occasionally—or skating in the winter, when the country was frost bound.*

The isolation of the OW was not usually a problem for Eula. In April 1897, however, a threatened "Indian uprising" proved how frightening it could be. When news of unrest on the nearby Northern Cheyenne Reservation reached the OW Ranch, Eula—pregnant with her first child—was encouraged to take shelter in Sheridan, traveling with two neighbor ladies:

I was not afraid and did not want to leave.... But Brother [Clarence] said he would not take the responsibility of allowing me to stay ... so I did as seemed best [and left]. I often think of the risk we took, driving into town, fifty miles across an unsettled country, not a ranch for twenty miles ... all alone with Indians "on the warpath...."

In 1909, the Kendricks relocated to Sheridan. The move not only allowed the children an opportunity to attend public school, but gave John Kendrick better access to the local business community, one in which he had become increasingly active over the years. It also offered Eula new opportunities. After nearly two decades of isolated rural living, she was ready to return to the active social life she'd known as a young girl in Greeley. She was soon busy with the Cecilian Club (a ladies' musical and literary society), the Methodist-Episcopal Church, and the Sheridan Women's Club.

When it was finished in 1913, her luxurious new home, Trail End, became the scene of teas, receptions, parties, dances, open houses, and dinners attended by businessmen, society matrons, politicians, ranchers, debutantes, and cowboys alike. An invitation to Trail End generally meant an evening of fine food, lively conversation, and dancing to the latest tunes.

With her husband's entry into state and national politics, Eula took on the role of political wife. An avid believer in his political views, she volunteered to handle all John's correspondence while he campaigned for office—no small task during a time when nearly all communications had to be handwritten or typed.

Once in the Governors' Mansion following John's victory in 1914, Eula proved to be a gracious hostess. Her finishing school education paid for itself with her ability to both entertain and be entertained. She gave credit for her social success to her mastery of the art of "small talk":

> *While I presided over the Governor's mansion, I learned that the social life is very essential. The hostess entertains all kinds and classes. Many … are strangers. The successful hostess, in order to lead in the conversation, and keep it flowing without constraint, must cultivate a line of light talk, carefully avoiding subjects of a personal nature, or that might offend those holding opposite opinions. In fact, she must be as diplomatic as her husband, and by her tact protect his interests* (Painter 1928).

For the two years she was in Cheyenne, Eula Kendrick was a busy first lady. She traveled throughout the state visiting her husband's constituents and performed a variety of ceremonial duties. She also entertained at the newly redecorated Governors' Mansion. In addition to regularly scheduled "at homes," or public open houses, Eula held an afternoon reception every Tuesday, gave a dinner or luncheon party once every two weeks, and hosted a myriad of special gatherings. During legislative sessions, her duties increased to include three dinners a week for legislators.

Eula's social graces were generally appreciated by the people of Wyoming and admired by the society editors of both local and statewide newspapers. Her scrapbooks were filled with clippings of her triumphs:

> *Governor and Mrs. John B. Kendrick were "At Home" last night at the executive mansion to upwards of seven hundred*

Eula Wulfjen Kendrick wearing inaugural gown,
Governors' Mansion, Cheyenne, 1915 (Moeller-Edwards Collection)

people who called upon the chief executive and his gracious wife in response to the invitation extended through the press. From eight until eleven there was a constant procession of arrivals and departures.

Not since the Governor's Mansion was built, has this beautiful house been the scene of a more brilliant social event than the one which took place in it yesterday—the "Cadet Reception," so perfectly planned and carried out by Governor and Mrs. Kendrick....

Eula Wulfjen Kendrick, about 1915 (Sykes, Chicago; Manville Kendrick Collection)

The executive mansion was agleam with brilliant lights last evening when the governor and his charming wife dispensed their delightful hospitality to a number of guests in the form of a dinner party....

Upon their arrival at the Governor's Mansion the visiting ladies from Colorado found a wide swinging portal, thru which they entered, and were received with cordial greetings and smiling faces, by Mrs. Kendrick, the wives of the state officers, and the legislative ladies....

John's unexpected election to the United States Senate in 1916 forced Eula and the rest of the family to move out of the Governors' Mansion and into the nation's capital. There, from an apartment on Sixteenth Street, Eula proved as popular a hostess and guest as she had been in Cheyenne. She and John, as well as Rosa-

Maye and Manville when they were in town, spent most of their evenings going to dinners, parties, receptions, theater openings, and concerts. Eula once again explained her popularity by referring to small talk:

> *Women in official circles are often criticized for indulging in chatter, but there is really a demand for an easy stream of small talk. Before meeting a dinner engagement, I always outline subjects for discussion that I think will be interesting to the partner who is chosen for me by my hostess* (Painter 1928).

Not all of Eula's time was spent indulging in idle chatter. In addition to her many social responsibilities, Eula was, like most ladies in her circle, involved in a variety of charitable activities. During the First World War, Eula wrote to the *Cheyenne Daily Leader* about the contributions Washington society was making to the war effort:

> *Nearly every one is contributing his bit, either at home or in clubs, meeting in church parlors, club buildings, or as we, the ladies of the senate, do, at the headquarters of the Red Cross.... It is here the boxes are packed and shipped to the front.... Even the girls are doing what they can.... Rosa-Maye joined a wireless class at Goucher and is very much pleased and interested in the work.*

Eula spent most of each year in Washington with her husband, but returned to Sheridan nearly every summer. Upon her return she was asked about what she had seen and done during her time away, particularly when her activities put her in close proximity to those prominently associated with national events. In 1918, for in-

Clipping from an unidentified Washington newspaper, 1927 (Julia Boyd, artist; Sheridan County Historical Society Collection)

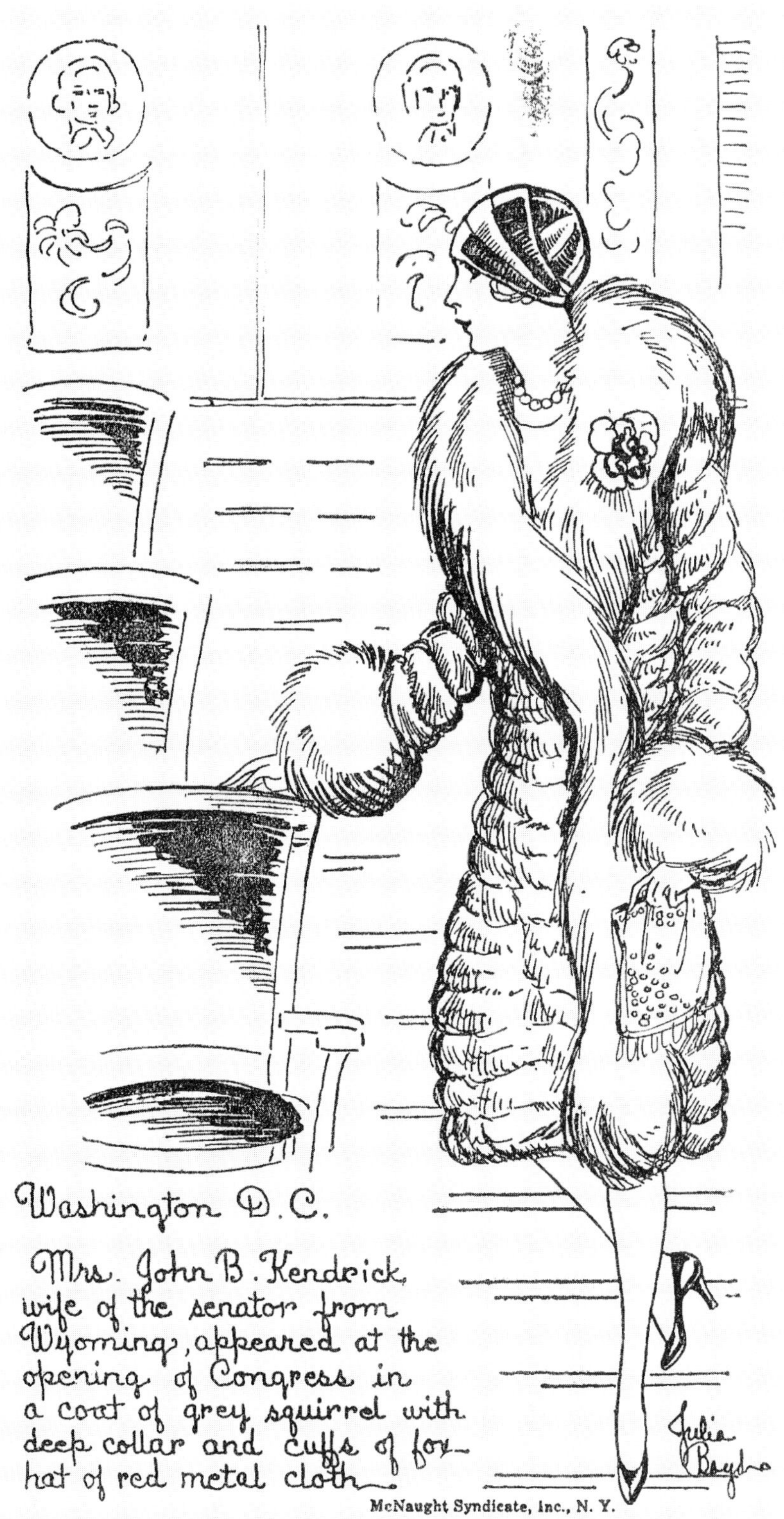

stance, the society editor of the *Sheridan Enterprise* posted a plea for Eula to share any up-to-date news she might have about the war in Europe:

> *Mrs. Kendrick has been a worker in the Senate Red Cross and gauze room of which Mrs. Marshall, wife of the vice president, is the leader. Mrs. Kendrick will have war news first hand. Dare we hope to hear about it later when she is rested?*

Though generally strong, Eula had frequent bouts of asthma and bronchitis. During her recovery, when doctors suggested she stay quiet, Eula worked on her favorite hobby: collecting the many clippings that chronicled the social, political, and charitable activities of herself and her family. Because of the "rags to riches" quality of their rise to fame, John and Eula were the darlings of the Washington press corps for several years. While magazine headlines referred to John as "The Sagebrush Senator" and "The Cowpuncher in Congress," Eula's portrait appeared in various women's magazines and frequent issues of the *Washington Herald.*

With her will to succeed, her strong desire for acceptance and her unerring eye for detail, Eula was able to go from ranch house to executive mansion with extraordinary efficiency. Along the way, she made both great friends and great enemies. One friend was author Frances Parkinson Keyes, who once described Eula in the following terms:

> *… a slender, sprightly, little lady, whose trim erect figure sets off to perfection frocks which are always the last word in smartness and elegance, and for which every accessory and adornment is always perfect; her soft, prematurely gray hair is always exquisitely dressed; and the perky, close fitting little*

hats which she affects frame a face that is always fresh, always animated.... (Keyes 1931, 14)

Not all observers were so generous when it came to the subject of Eula's clothes or her social motivations. In 1916, for example, Eula was soundly criticized by a writer at the *Saratoga Sun*:

[Kendrick's] good wife is so ambitious that when he was elected governor she could not find anything nice enough short of New York for a gown for the inaugural ball. Think of a woman pushing a man for social position, and when her husband became governor of this great state that she was too good to wear a dress that she could procure in the West, not even Cheyenne or Denver, but traveled to New York City and paid something like $10,000 for the gown. Did that make Kendrick a better governor for you and I or did it give us the hint of why Kendrick purchased the governor's office.

Though the story of the $10,000 gown was inaccurate (it was indeed purchased in New York, but at a wholesale house and at a wholesale price), it is true that Eula Kendrick was no mere helpmate or "power behind the throne." She wanted a measure of fame for herself, and her husband's political office helped her attain that fame. Aside from her position in society, Eula occupied a spot in Washington's feminine political world. She was a charter member and treasurer of the Women's National Democratic Club and was an active member of the Congressional Club as well. In 1923, Eula served as acting president of the Senatorial Women's Club, also called "The Ladies of the Senate," a position traditionally held only by the wife of the vice president of the United States. The vacancy arose when Grace Coolidge followed her husband to the

White House after President Warren G. Harding's sudden death. Eula noted in an interview with the *Austin American* that the Congressional Club was a useful source of information relating to the issues faced by her husband:

> *Mrs. Kendrick had been attracted to this organization very strongly, on her arrival in Washington, for it presented to the women who make up the senatorial families the same opportunity to know each other and to get the viewpoint of different sections, which is enjoyed by the men in their daily association with each other.*

In 1933, at the relatively young age of sixty-one, Eula suddenly became a widow and her entire world was turned upside down. After sixteen years of energetic involvement in Washington's social and political circles and nearly two decades of being sought out by the press, Eula found herself alone and ignored. Despite her apparent popularity, John's death resulted in the end of Eula's active role in Washington society. Because the reporters and photographers were no longer interested, the public's concern with

Three first ladies of Wyoming: Nellie Tayloe Ross, Julia David Carey, and Eula Wulfjen Kendrick (John B. Kendrick II Collection)

her activities waned as well and the impressive stack of clippings ceased to grow.

One of the last public tributes Eula received was in 1934 when, during a Congressional memorial service in the House of Representatives, Senator Robert D. Carey of Wyoming acknowledged her contributions to the success of Senator Kendrick's career:

> *How much inspiration for Senator Kendrick's knowledge and success resulted from the influence of that splendid lady, I shall not pretend to say; but having known John B. Kendrick for more than 20 years and appreciating the value he placed upon the finer things of life, my impression is that he gave to her the greater credit for all the happiness and success that had come to him* (United States Congress 1934, 57–58).

Eula returned to Sheridan and for twenty-five years lived at Trail End with her son and his family. She worked on her scrapbooks, made notes for memoirs that were never published, and

Eula Wulfjen Kendrick and Governor Leslie Miller at dedication of Kendrick memorial in Sheridan's Kendrick Park, 1936 (Harmon Collection)

fiercely guarded her husband's political and personal memory. She took steps to ensure that John Kendrick's political legacy would be untarnished by any personal or professional scandal: while she later donated his political papers, daily appointment books and business diaries to the University of Wyoming, Kendrick family lore states that after his death, Eula burned all of John's personal diaries.

Eula Wulfjen Kendrick, about 1928 (Underwood and Underwood, Washington, D.C.; Hoff Collection)

A few years before her own death, Eula moved to San Antonio to live with her daughter. She died in her native state of Texas in June 1961 and is buried next to her husband in the Sheridan Municipal Cemetery. Shortly after her death was announced, the editor of the *Sheridan Press* made these comments:

> *The visions of dusty cattle trails and tired cattlemen and cowboys are … seen through memory's eyes. Too often perhaps it has been pictured a man's history, these early days in the Sheridan area, but it was a woman's history too. Women hearty in nature and heart were here to work beside and to provide the necessities of home to men who spent their days among the cattle herds. Yes, women were making history, too. And Mrs. Kendrick was one of the history makers.*

As the daughter, wife, sister, and mother of pioneering cattlemen, Eula Wulfjen Kendrick possessed both strength and determination. Forged by the hot sun of the Wyoming prairie, these characteristics combined to create a force in Eula that few were able to ignore.

The Best Cowhand on the Ranch

ROSA-MAYE KENDRICK

Rosa-Maye Kendrick, 1917 (Harris and Ewing, Washington, D.C.; Manville Kendrick Collection)

Six years into their marriage, after several false alarms and at least one possible miscarriage, John and Eula Kendrick finally became parents. Since there was no hospital in Sheridan at that time, most women delivered their children at home, relying on midwives to provide any needed medical assistance. Eula Kendrick, however, was not willing to take chances. In 1897, shortly before the impending birth, Eula left the OW Ranch and went to Sheridan to become a patient of the "Cutter sisters," two Canadian nurses who received patients, usually pregnant ranch wives, in their home. In her later years, Eula wrote:

> *I went again into town late in June or early in July, to await the arrival of the "stork"—long delayed in visiting us. Mr. Kendrick left me at the [Sheridan] Inn, where I spent a week or ten days, then went out to my nurse's home. The little daughter arrived in due time assisted by Dr. M. A. Newell and Miss Anna [Cutter] and hovered over by Mr. and Mrs.*

A. S. Burrows, our partner in the Kendrick-Burrows Bank and close friends of ours.

The baby girl (who had always been referred to as "he" in John Kendrick's letters to his wife during her pregnancy) was given the name Rosa-Maye: "Rosa" in honor of John Kendrick's younger sister (a well-educated teacher who had died of typhoid in 1893), and "Maye," his mother's maiden name.

When she was five weeks old, Rosa-Maye was introduced to the OW Ranch where she lived for the next twelve years. Isolated as they were with no school or other children nearby, Rosa-Maye and her brother Manville developed a close relationship with their mother, who was also their teacher:

Rosa-Maye Kendrick and her mother, 1897 (Huffman Studio, Miles City; Manville Kendrick Collection)

> *Not in studies alone was Mrs. Kendrick her children's mentor, but the young mother taught them to ride, swim, skate, and shoot, and by this close companionship attained the enviable position of being a pal with her children* (Painter 1928, 201).

Rosa-Maye had a fairly typical ranch childhood. She came to town regularly, visited family in Texas and Colorado, and like most girls her age, developed romantic "crushes," usually on cowboys from neighboring ranches. She was happy living at the OW Ranch and did not particularly want her life to change. Even after the family moved to Sheridan in 1909, staying first at the Sheridan Inn and later in Trail End's carriage house, the ranch was still home to her.

Rosa-Maye Kendrick, 1900 (Rinehart Studio, Omaha; Manville Kendrick Collection)

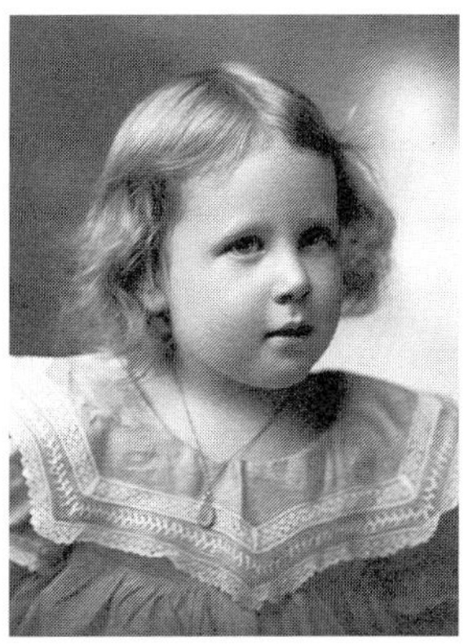

After moving to Trail End in 1913, "home" was no longer what she'd known it to be. Instead of a cozy ranch house, she now lived in a grand mansion. Not only that, there were bigger changes in the air, most of them related to John Kendrick's growing interest in politics. Rosa-Maye became depressed, worrying about what would happen if her father succeeded in his political pursuits. She wrote in her diary:

> 31 July 1913: *They spoke about Dad selling the Ranch at the table today. Such conversation is more than I can stand. Makes me want to crawl away and die. Guess it's only a question of time until he sells the Ranch anyway. Uncle is sure he is going on Cheyenne Trip. Spoke about it yesterday. Guess we'll go too. Wish I was "home" tonight....*

Not one to pout, however, Rosa-Maye was determined to make the best of what she saw as a bad situation. After spending a little more time in town, the teenager grew to appreciate its advantages and opportunities, particularly the social ones. Diary entries less than a year later show her change in attitude:

> 12 March 1914: *Harry [Kaye], Brother and I are taking new dances at Central Hall Wednesday nights and are learning fast. Expect to go to a big dance Sat. night.... Last night twelve girls who have not homes large enough to entertain in gave a dance at the I.O.O.F. Hall.... The boys are now talking of giving a similar dance in honor of the girls.*

After moving to Sheridan, Rosa-Maye attended classes at Taylor and Central Schools. Although she was allowed to come back to Sheridan to graduate with her Sheridan High School class in the spring of 1915, Rosa-Maye received the bulk of her high school–level education at Ely Court in Greenwich, Connecticut, which she entered immediately following her father's successful bid for the governorship in November 1914.

Rosa-Maye Kendrick in Sheridan High School graduation gown, 1915 (Fuller Studio; Hoff Collection)

While at Ely Court, she completed a college preparatory course designed to make up for deficiencies in her earlier education, primarily as related to math and science. Though she had compared favorably with her classmates in the Sheridan schools, she was considerably behind in what Ely Court administrator Elizabeth D. Ely referred to as "foundation work":

> *If [Rosa-Maye] finishes her college preparatory course we will give her our diploma, but we are sorry to find that considerable foundation work has been neglected, and it is a stiff piece of work to fill up the crevices, but we are working on it.... Both we and she feel somewhat oppressed by all that is to be done before June, but we are all working at the problem with all our energies, and hope to accomplish it.*

Rosa-Maye was as determined to go to college as her parents were to have her go. Just as her brother had been enrolled in Harvard from an early age, Rosa-Maye was enrolled in Vassar. The prestigious women's college was initially willing to accept her home-schooling and her classes from Sheridan High School, but following her entry into Ely Court, Vassar dropped Rosa-Maye from its rolls, citing Ely's lack of accreditation. Although she was eventually reinstated, Rosa-Maye took the cut personally and stated that if Vassar were the last college on earth, she would not go there. After considering a variety of other institutions, Rosa-Maye eventually decided on Goucher College in Baltimore, Maryland.

Although not the most academically gifted student, Rosa-Maye was always popular and very active on campus. She performed in theatrical productions and wrote for the student yearbook. Her diaries and scrapbooks are filled with the many details of her busy social life. Following a 1915 trip to the Percy Rockefeller estate, where she rode thoroughbred horses and had luncheon with the Rockefeller family, Rosa-Maye wrote a letter to her father in which she expressed her appreciation for the opportunities she had been given:

> *I know now, even if I thought I knew before, what a wonderful opportunity it is for a girl to come East to school. She doesn't realize how narrow and prejudiced she is until she comes in contact with the different atmosphere. If then she decides, as of course I do, that her own West is God's own country, all the better, for she has seen both sides of life.*

During her time at school, Rosa-Maye did indeed see a different side of life. With a number of military academies nearby, the denizens of all-female schools such as Ely Court and Goucher College were never lacking for adequate male companionship. Cadets frequently acted as escorts at the balls, teas, and receptions held for Rosa-Maye and the other young ladies in the social circles of Greenwich, Baltimore, and Washington. In late 1919, through the efforts of her cadet friends, Rosa-Maye was chosen to christen a new ship for the United States Merchant Marine fleet. When invited to become the sponsor of the vessel, she first thought to name it after her hometown of Sheridan. When she discovered that name had already been used, she settled on the name *Sundance.* The christening was held at the Hog Island Shipyards in Baltimore in October 1919:

Rosa-Maye Kendrick at launching of SS Sundance, *1919 (Sheridan County Historical Society Collection)*

> *. . . the party proceeded to the huge way, upon which the* Sundance *was awaiting its christening. The platform surrounding the bow of the vessel was decorated with bunting and American flags and just before the word for the launching was given, Miss Kendrick was presented with a beautiful bouquet of American beauty roses. . . . Miss Kendrick delivered a vigorous blow with the christening bottle just after [the ship] began to move and before it got out of reach.*

Following her graduation from Goucher College in 1920, Rosa-Maye experienced one of the grand institutions of the American upper class: her "coming out." Beginning with a fancy dress ball, this formal introduction to society lasted for a year and was a way for young ladies to circulate, meet eligible bachelors, and consider proposals of marriage—all under the watchful eyes of respectable female guardians.

The typical debutante married shortly after her debut. Rosa-Maye, however, did not. Neither did she pursue a career, although she did work for some time in her father's Senate office chambers, learning about the issues in which he was interested. Like other well-to-do Washington ladies, Rosa-Maye occupied her time by lunching with friends, attending receptions and openings, and involving herself with a variety of charitable organizations. She kept track of her schedule in her diary:

> 18 January 1925: *Went down to the library to do my Junior League work. Shellacking Braille sheets for the Blind.... Afterwards attempting a literary career in the form of an essay. Came home in time to rest before dinner. Went afterward to the Diplomatic reception.*

Rosa-Maye's essays were not written solely for her own amusement. Several of them were published by Washington newspapers. Pertaining as they did to "western" issues such as irrigation, land use, and range practices, most of these articles reflected her father's opinions on the issues (making them useful campaign tools when reprinted in western papers). Rosa-Maye also wrote articles about the sights and sounds of her Washington life, several of which were printed by Wyoming papers including the *Laramie Republican-Boomerang* and the *Sheridan Enterprise.* When her series of articles was over in 1926, the editor of the *Boomerang* offered these words of thanks:

> *Miss Kendrick writes [in] such an altogether charming and delightful style that it is to be hoped that she will continue in this line of work. Very few women, in our observation, even the much heralded feature writers of some of the larger and*

more prominent magazines, have the ability to impart their information in such a readable style as does Miss Kendrick. She has indeed placed a large number of Wyoming readers greatly in her debt.

Rosa-Maye Kendrick (right) and friend on roundup, about 1922 (Hoff Collection)

More than anything during this time in her life, Rosa-Maye looked forward to her time at Trail End and the OW Ranch. Although she made several trips to Europe, Cuba, and Panama, summers were almost always spent in Wyoming and Montana. When she was not in Sheridan taking care of her invalid grandmother, accompanying her mother on endless rounds of social calls, or playing golf on the south lawn of Trail End, she was at the OW for roundups and short cattle drives. As Eula Kendrick noted in her memoirs, Rosa-Maye not only loved ranch work, she was good at it:

The boys all said [Rosa-Maye] was always the best cowhand on the ranch. She was at least the happiest and most interested and would come home from weeks on the round up full of anecdotes of honest human incidents of both man and beast—or to her, man, horse and cow were equally human.

The OW Ranch became a place for young people to escape from the social whirl of Sheridan and Washington. Along with Manville and her cousin, Eula Williams, Rosa-Maye played host to friends and family who wanted to experience the simplicity of OW life:

Rosa-Maye Kendrick (left) and cousin Eula Williams, 1920 (Hoff Collection)

3 August 1925: *Harry [Kaye], Dotty [Mondell] & I loaded Brother's Dodge from beds to chewing gum & started for the ranch on pleasure bent with a smile & wave for everyone. Arrived about 5:30. Began collecting beds. Spread out and raised my Teepee. At supper amazed at the quantities of food Harry & Dotty consumed.*

Rosa-Maye spent a great deal of time campaigning for her father and other Democratic candidates. In September 1926, Senator Kendrick agreed to make a campaign swing through western and southwestern Wyoming in support of Democratic Governor Nellie Tayloe Ross, then running for reelection. Rosa-Maye was all set to accompany John on the journey, which included a two-day train trip through Idaho. In a moment of whimsy and despite her self-admitted lack of experience with long-distance driving, Rosa-Maye impulsively volunteered to be her father's driver. His acceptance of her offer was equally unexpected and what followed was an adventure for both father and daughter.

Stops on the campaign trail included Ten Sleep, Lander, Pinedale, Jackson, Kemmerer, Afton, Evanston, Green River, Thermopolis, Basin, Greybull, and Deaver. Along the way Rosa-Maye had ample opportunity to hear her father's dreams for his adopted state. In an unpublished manuscript entitled "Fragments of My Trip with Daddy," she recalled some of those dreams:

> *Father had a vision which transformed the rolling sagebrush prairies over which we rose and dipped for mile on mile, into pleasant acres dotted with farmhouses, burdened with crops, and threaded by canals bearing the miracle-working waters. He was confronted now by the difficult task of making this dream a reality.*

Through muddy roads, substandard lodging, and the occasional inhospitable audience, the pair continued on uncomplainingly. Senator Kendrick had a job to do, and Rosa-Maye was willing to put up with almost any hardship in order to spend time with her father.

Though she loved her mother and brother, Rosa-Maye was definitely "Daddy's little girl." She spent as much time with him as possible and when they were apart, wrote long letters describing what she was doing and with whom. On June 14, 1920, when she sailed to Europe for the first time, she affectionately expressed her beliefs concerning a father's purpose in life:

> *You may wonder why Daddys are, when you come home from a hard day's work and see a frivolous daughter just going out for the evening. Still I assure you that daughters aren't as frivolous as they seem and Daddys are to remind*

them that there are other things in the world besides play. Most of all they set standards (at least for this little daughter) by which all other men are judged; perhaps, this is their best "raison d'être."

Whether it was the high standards embodied by her father or a perceived lack of qualified candidates, Rosa-Maye seemed in no rush to get married. She had dated steadily since her mid-teens and had even entertained a marriage proposal or two. But it was not until she was nearly thirty years old that she agreed to marry a man who had courted her steadily for five long years.

Major Hubert Reilly Harmon, 1927 (Vandyk, London; Sheridan County Historical Society Collection)

Twenty-seven-year-old Hubert Reilly Harmon was a major in the U.S. Army when he first met Rosa-Maye Kendrick at the wedding of mutual friends in 1922. He quickly became part of her social circle, taking her riding and escorting her to a variety of social functions. As the weeks went by, Hubert was featured more and more prominently in Rosa-Maye's diary. After only a month, Rosa-Maye noted in her diary that Harmon "became serious in my direction." But, as she was dating another man, she put him off, at least temporarily.

Hubert was both a diplomat and a flyer who used all his skills to maneuver himself into a future with Rosa-Maye. Born into a military family in 1895, Pennsylvanian Hubert Harmon was more than up to the challenges of courtship. A 1915 graduate of West Point (in the same class as Generals Dwight D. Eisenhower and Omar Bradley), Harmon had served in both France and Germany during World War I. By 1922, he was serving as a military aide at the White House. On April 1 of that year, less than two months after their first meeting, Hubert risked his relationship with Rosa-Maye by proposing. Two days after the failed attempt, he sent orchids and a "beautiful incoherent" letter that, in her eyes, "brought back sunshine and the spring."

Rosa-Maye Kendrick Harmon, 1927 (Hoff Collection)

The Kendrick-Harmon courtship was a turbulent one: he asked her to marry him dozens of times, both in Washington and in Wyoming, before she finally accepted his proposal in late 1926. Though they'd been an "item" for nearly five years, their engagement was a surprise, at least to the society editor of the *Cheyenne Tribune*:

> *The engagement of Miss Rosa-Maye Kendrick … to Major Hubert Harmon of Washington, D.C., was announced today. … Word of the engagement came as a surprise to Washington society, where Miss Kendrick is a great favorite.… Miss Kendrick and Major Harmon had been friends for years and for the past three years the Major has been a summer guest of the Kendricks, but rumors of an engagement have been denied.*

Shortly after their 1927 Washington wedding at All Soul's Church (attended by Mrs. Calvin Coolidge and a bevy of political dignitaries), the Harmons moved to London where the major was assigned to the post of military air attaché at the American Embassy.

As the daughter of an influential American politician married to a handsome young military officer, Rosa-Maye had many doors opened to her. She and Hubert toured the countryside, visiting castles, going to balls, and taking in local fairs and events. A collection of her chatty letters home was eventually published, without her knowledge, by Eula Kendrick, who felt that "unstudied as they are … they give us a new version, in a new way, of the old customs of an old country" (Harmon 1928, preface).

One old custom described by Rosa-Maye was her presentation to the king and queen of England. This was her second appearance at the Royal Court; the first was while in the company of her mother in 1924. For that event she had a special gown created to conform to Court regulations. For this 1927 presentation, Rosa-Maye simply altered her slightly used wedding gown:

> *My preparations for court were simple. I took the court train of my wedding gown to a dressmaker, patronized by the American ladies of the Embassy, to have shortened (the length required must be eighteen inches from the heel) then the dress, feathers and veil to the cleaners and I was ready, except to devise some means of applying the feathers and veil to my short hair without the medium of a bandeau, as I had done before in 1924. I had definitely decided that I would not compete with the diamond tiaras this time, except as a last resort* (Harmon 1928, 49).

Other English adventures included the races at Ascot, the Royal Opera at Covent Garden, a royal garden party at Buckingham

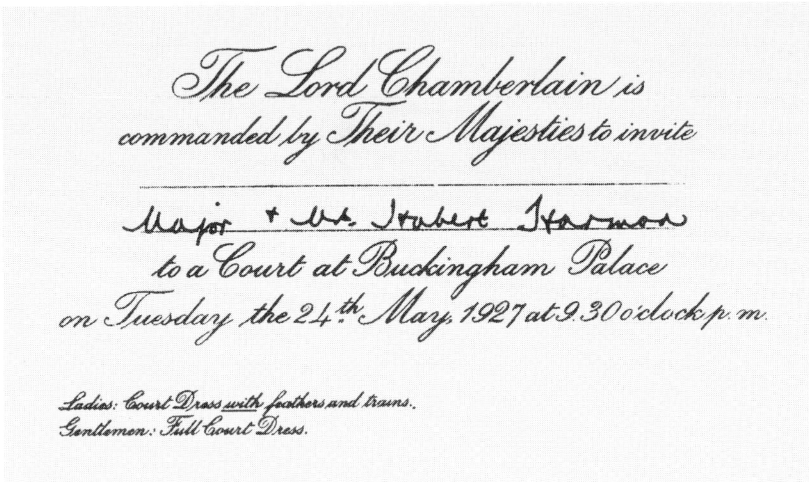

The Lord Chamberlain is
commanded by Their Majesties to invite
Major + Mrs. Hubert Harmon
to a Court at Buckingham Palace
on Tuesday the 24th May, 1927 at 9.30 o'clock p.m.

Ladies: Court Dress with feathers and trains.
Gentlemen: Full Court Dress.

Invitation to Buckingham Palace, 1927 (Trail End Collection)

Palace, and Charles Lindbergh's arrival in London following his historic solo flight across the Atlantic:

> *At first we could only distinguish a faint blur. Gradually it developed into the escort of honor, perhaps seven or eight of the British Airways Ships, flying in formation. When at last [Lindbergh] was visible, (a tiny gnat amongst huge black beetles) we recognized his plane by its single wing. Nearer and lower they came, and at last swept over us with a thundering roar. The crowd went wild* (Harmon 1928, 64).

After returning to the United States a few years later, the Harmons were posted at the U.S. Military Academy in West Point, New York. Following Hubert's promotion to the rank of general, the family moved to Randolph Field in Texas where he commanded the Gulf Coast Air Training Center. During the course of his career they also lived on military bases in California, Georgia, and Kansas. As the capstone to his active military career, General Harmon served as the first superintendent of the U.S. Air Force Academy.

Though living far away, Rosa-Maye never lost her love for the West. She and her two children, Kendrick and Eula, visited Trail

President Dwight D. Eisenhower (in dark suit and hat) escorted by General Hubert Reilly Harmon, about 1953 (Kendrick Cattle Company Collection)

End and the Kendrick ranches often while Hubert was stationed out of the country. Although he hoped otherwise, her father was well aware that Rosa-Maye would never again live full-time in Wyoming, either at Trail End or at any of the ranches. As John Kendrick noted in a 1932 letter to Manville:

> *... your sister will never find it possible to [live here]. An army officer is not unlike a Methodist preacher who has neither a home nor even a native state, but is constantly moving from place to place until the age of retirement.*

Despite her physical absence, Rosa-Maye maintained an abiding interest in the family business. In hundreds of letters to her father, mother, and brother, sent from wherever she and Hubert were living, she never failed to mention some aspect of ranch business. Although others managed the cattle operation, she always made her thoughts known. Even after Hubert's death in

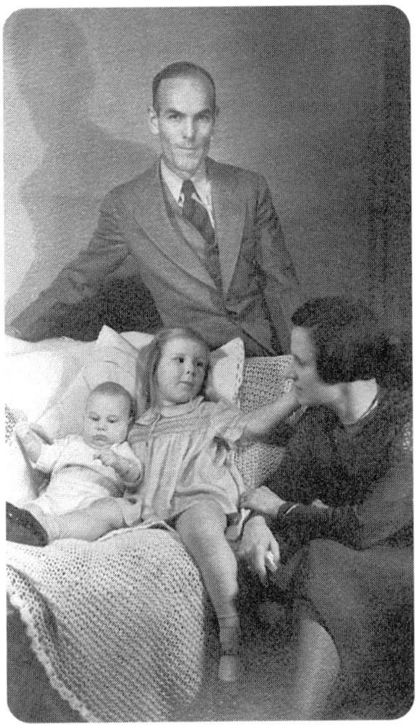

Far left: Hubert Harmon (standing) with Kendrick, Eula, and Rosa-Maye Harmon, 1934 (Sheridan County Historical Society Collection)

Rosa-Maye Kendrick Harmon, 1928 (Harris and Ewing, Washington, D.C.; Moeller-Edwards Collection)

1957, Rosa-Maye did not return to Wyoming. She stayed instead in San Antonio, Texas, where she died in 1979 at the age of eighty-one. She is buried alongside her husband on the grounds of the Air Force Academy in Colorado Springs.

In 1923, Dr. J. Youge Barnes of the Hotel Lafayette in Washington, D.C., prepared a "constructive analysis" of Rosa-Maye's personality and disposition, which seemed to accurately summarize the forces that drove her throughout her life:

> *A positive temperamental type, meaning a type with strong determinations and the power in mentality to plan, devise and promote.... This positive nature is well aligned and is followed by strong loyalty, devotion and sacrifice with affection, sympathy, sentiment and emotion.... Impatient.... Impetuous.... Strong sense of integrity and honor and with it ability to bear and carry responsibilities.*

In the Footsteps of His Father

MANVILLE KENDRICK

Manville Kendrick, 1918 (Cunningham Studio, Exeter; Manville Kendrick Collection)

Brought up as he was on the OW Ranch, Manville Kendrick might have stayed there, married a local girl and spent his years working the land the way most young men in the area did. But with Senator Kendrick for a father and Eula Wulfjen for a mother—both determined individuals with their eyes on the future—Manville was destined to lead more than the simple ranch life. He received a university education, was introduced to culture and the arts, and eventually became an influential rancher and businessman in his own right.

Manville was born on July 20, 1900, in a third-floor room of the Sheridan Inn. He was named for Hiram S. Manville, a co-founder of the original OW outfit and one of John Kendrick's early mentors. According to Manville Kendrick, in a 1971 letter to Anna Hoblit:

> *H. S. Manville was not related to my father. Having been orphaned at an early age, my father regarded Mr. Manville*

(who was considerably older) in loco parentis. *As a result, I was named for Mr. Manville.*

Hiram S. Manville with namesake Manville Kendrick, 1900 (Rinehart Studio, Omaha; Manville Kendrick Collection)

Manville's early years were spent on the family ranch and later at Trail End. After his father was elected to the governor's office in 1914, Manville went to school briefly in Cheyenne where he was a member of the high school cadet corps. In the fall of 1915 he was sent to Phillips-Exeter Academy, a prep school in New Hampshire. Although he was an ambitious man who wanted his only son to have the best education possible, it was not easy for John Kendrick to send Manville so far away. He wrote to a friend in 1914:

> *While I have not yet given up the hope of having Manville finish his education at Harvard, I have been inclined to hope that he might make the necessary preparation in the West, or at least shorten up the time that would be required at preparatory school. With just one boy in the house and very few opportunities to be with him even when he is at home, it seems to be next to impossible for me to send him away to school until he is fairly well grown.... [But] we shall endeavor to make some arrangement for the boy to go to Harvard.*

Manville Kendrick, about 1908 (Kirkland Studio, Denver; Manville Kendrick Collection)

And what did Manville think of the prospect of all this quality education? In 1915, during a visit to Phillips-Exeter, Governor Kendrick found the following note his son had written to himself: "I will get out of here in two years or bust! Watch me!"

Perhaps because of his discontent at being so far from home, Manville did not excel in his academic pursuits.

He was at best an average student, his highest scores being in literature and languages (Greek, Latin, French, and German). Like his sister, some of his preparatory work had been less than adequate; throughout his academic career, for example, he failed miserably in mathematics. Nevertheless, by the summer of 1918, Manville had graduated from Phillips-Exeter and was bound for Harvard College in Cambridge, Massachusetts.

Despite his fairly simple upbringing, or perhaps because of it, Manville developed expensive tastes during his time at school. Along with his wealthy eastern classmates, he followed the latest trends and fads: raccoon coats, tuxedos, Egyptian cigarettes, Broadway shows, and fast cars. All of this was costly, and Manville's constant need for spending money was not lost on his father. In a 1919 letter, the Senator teased his son:

> *I had a birthday a few days ago and received as a present the enclosed one dollar bill from your grandfather. I have since felt a great deal of anxiety as to what was best to do with this money, and have written your grandfather to that effect. I explained to him that I felt the urgent need of keeping the money in the family and so far as I knew you were the only member of the family who never expended any of your own funds and so on this account I thought I would send it to you.*

On October 15, 1918, shortly after his arrival at Harvard, Manville was selected for immediate military service and ordered to report to Division 4, Cambridge, Massachusetts, to begin active duty. He was assigned to Company D of Harvard's Student Army Training Corps. War was still raging in Europe and Manville was eager to do his part. Long interested in airplanes, Manville entered the military with the aim of joining the Air Service. On September

25, 1918, just days after registering for the selective service, Manville sent his father the following telegram: "Aviation Corps open for short time only … if accepted would enter officers training camp immediately … may I try … wire immediately."

While John and Eula Kendrick were not opposed to the idea of Manville serving his country—indeed, they applauded his patriotism—they were appalled by the idea of their only son joining one of the most dangerous branches of the military. Rather than voice an emotional appeal, John tried to reason with Manville. A few days after receiving the telegram, John wrote to his son:

> *If a man wants just to do something sensational and quit, why undoubtedly your best plan would be to go into aviation, but certainly if you chose that branch of the service with a view of keeping it up, you are not expecting to become an important man in your State and community in the next twenty or thirty years, because aviation doesn't lead in that direction.*

Undeterred by that argument, Manville wrote to Senator Francis E. Warren, his father's colleague and one-time political rival, asking him to write a letter of recommendation to the Army Air Service. Warren responded on October 4 with a letter to the Chief of the Division of Military Aeronautics:

> *[Manville Kendrick] … is a splendid specimen of young manhood with a most pleasing personality and appearance which, we cannot doubt, mean much in a young man's favor when he is under consideration for training as an officer.… I beg to endorse and ask your consideration of the application referred to … I feel that young Kendrick will be a success in your Service.*

That same day, Warren wrote to Manville about his prospects:

Manville Kendrick, member of the Student Army Training Corps at Harvard, 1918 (Towles, Washington, D.C.; Manville Kendrick Collection)

... I commend you for your "high" ambition and hope you may qualify for the Air Service as you desire.... I can understand your Father's feelings in giving his consent to your application rather involuntarily. But I know that in his heart he wants you to serve in the place where your inclination leads you, for that is where you will have every incentive to "make good...."

Unfortunately for Manville, such was not his father's attitude. Instead, Senator Kendrick wrote to Warren and asked him to put off mailing the recommendation until he made one final plea to Manville's common sense. This was done and by October 5, father and son arrived at a compromise: if Manville would stay in school the rest of his freshman year, he could join the Air Service in the spring—if he still wanted to:

> *By deferring your application you will have the benefit of the military training as well as one year's academic work in college. Such a course will please your mother and myself beyond measure, and as stated before we would, whether willingly or not, cheerfully cooperate with you at that time in carrying out any plan upon which you have your heart set.*

Throughout the rest of the school year, Manville remained in the Student Army Training Corps, receiving one-third additional academic credit because of his military status. The issue of his enlistment in the Air Service was soon moot, however, as the war was over by the middle of November 1918.

Despite his mediocre grades, a serious bout with the Spanish Influenza, and the distractions of the Boston and New York night life, Manville graduated from Harvard with a bachelor's degree in government. He then went on to attend Ames Agricultural College in Iowa where he learned many of the skills that would help him later when he began managing the Kendrick Cattle Company.

Manville Kendrick at OW Ranch, 1926 (Hoff Collection)

In the summer of 1922, shortly after his graduation from Harvard, Manville went to work at the Kendrick ranches. In a letter to the ranch manager, Manville's uncle Clarence Wulfjen, John Kendrick outlined what he hoped would be "the boy's" course of study while under Wulfjen's tutelage:

> *... his training from now on should be calculated to give him a general grasp of the business.... I don't want to see the time come when you are separated from the management of the business but I would like to see this boy fall into your footsteps and help you just as much as you care to load it on to him. If he takes to the business he will in the near future commence to suggest things to you that will be helpful in the management and that will lighten your burdens every day ... he will be a pillar of strength to you.*

Kendrick also seemed deeply concerned about Manville's physical well-being. Although never strong (suffering throughout his life with a variety of stomach disorders and other ailments), Manville was nevertheless eager to work outdoors on the difficult and sometimes dangerous tasks associated with ranch life. It was these risks that the Senator hoped to have his son avoid:

> *Because of what it all means to his mother and to myself, I would like to have you encourage him incidentally, not too*

much, to avoid taking unnecessary risks of his life and limb. He must of course take his place with other men and do as other men do, and I have no disposition to avoid that, but I would feel discouraged to consider for him anything like my own record of risks, and you happen to know that this record has been extensive.... I know too well your own anxiety to lead him right and to protect him at every turn you can from unnecessary risks and hazards.

Manville managed to survive the hazards of ranching and even had time for a social life. In between roundups, shipping, and haying, he had several opportunities to visit his parents in Washington, D.C. There he met a young debutante named Clara Diana Cumming. Diana was the daughter of U.S. Surgeon General Hugh Smith Cumming and his wife Lucy Booth Cumming, both members of old Southern families of considerable social standing.

Diana Cumming, 1922 (Bachrach, Washington, D.C.; Manville Kendrick Collection)

During her early childhood, Diana and her brother Hugh lived the gypsy life of the government employee's family. She was born in 1901 at Blackbeard Island, Georgia, where her father worked at a Public Health Service quarantine station. They later moved to a similar facility on the opposite side of the continent, at Treasure Island near San Francisco. Just days before the San Francisco earthquake of 1906, they moved to a U.S. Public Health Station in Yokohama, Japan.

In the mid-1910s, Hugh Cumming moved his family to Washington, D.C., and enrolled his children in school. While young Hugh Cumming Jr. was sent to the Virginia Military Institute,

Diana attended Western High School. She proved to be a popular student, excelling in riding and shooting—she was the only female to compete on the school's rifle team. After graduation, Diana attended the Cathedral Girls School, where she served as president of her class.

Prior to dating Manville, Diana had been squired about almost exclusively by cadets from West Point, Annapolis, and the Virginia Military Institute. Therefore, her engagement to the young Wyoming rancher came as somewhat of a surprise. In fact, Diana had known Manville Kendrick for several years: in 1923, her first cousin had married Manville's first cousin. Plus, the Cumming family lived on the same street in Washington as the Kendricks, making Manville almost the boy next door.

The romance between this blue-blooded debutante and her Harvard-educated cowboy was tailor-made for publicity. The

Diana Cumming sitting atop bull at LX Bar Ranch (Manville Kendrick Collection)

courtship delighted society editors who were always looking for fresh faces and interesting stories to put before a gossip-hungry public. Frequent accounts in the Washington and New York newspapers detailed Manville's extended visits to Washington and Diana's fun-filled excursions to Wyoming.

In 1929, in the Bethlehem Chapel of Washington's National Cathedral, Manville and Diana were married. One of the biggest society weddings of the year, the event was attended by First Lady Grace Coolidge, the Secretaries of State, Treasury, Navy, Agriculture and Labor, dozens of members of Congress, plus ambassadors and foreign ministers from Asia and Central America. Following a honeymoon cruise from Baltimore to San Francisco via the Panama Canal and the West Indies, Manville and Diana moved to Sheridan.

Although they lived there for nearly thirty-two years, the young couple never wanted to make Trail End their permanent home. Diana in particular wished to have a home of her own, away from the frequently conflicting demands of her husband and his mother. In May 1931, Senator Kendrick professed to offer not a "single objection to the plan," saying it was all Eula's doing. In his letters, however, he went out of his way to make living at Trail End an attractive proposition:

> *... my only interest in making recommendations is to aid you in avoiding mistakes.... It ought to be more economical for you to live with us ... than in your own home. Moreover, it ought to be possible for Diana to get away any time she wishes, if we are in the house. Even if you moved into your own home she probably would not feel anything like as safe to leave the baby and go with you as she would if he were*

looked after by a nurse and the nurse were looked after by his grandmother.

Three years earlier, before the wedding had even taken place, it was Eula Kendrick who had supported the young couple in their efforts to strike out on their own. She wrote to Manville:

I agree with you in that Diana should have her own home if we wish to make her happy in this transplanting process.... With her own home either over the garage or in a cottage across the street on the north ... she can fill her time with attention to the little homey details all brides love and can entertain her friends at will, which after all is the only form of amusement Sheridan offers. You may be sure I will aid you in every way possible as will Father. He may not meet your suggestions at first, but I feel confident he will come to them if you present them with a good sound argument and let them sink in.

Even though they purchased land south of Sheridan and had detailed house plans drawn up, in the end Manville and Diana abandoned such dreams and moved into Trail End permanently, making their own "apartment" out of the former guest wing. Here they raised their two children, John Benjamin Kendrick II and Hugh Smith Cumming Kendrick. When tensions ran high between Diana and Eula, either one or the other (and sometimes both) would make a getaway to one of the ranches, Washington, D.C., or some other location. When Manville was asked in a 1987 interview how he coexisted with two such strong personalities, he simply said, "I keep out of the way."

While Senator Kendrick was serving his country in Washington, Manville gradually became more involved in managing the

Manville Kendrick with sons John and Hugh at the carriage house, about 1939 (Hoff Collection)

family ranches. Making the transition from Harvard to hayfield was not always easy, but he was a quick learner, rarely having to be told something twice. There were, of course, problems. In December 1922, for example, Manville learned from his father (via a letter to Clarence Wulfjen) the importance of attending to details:

> *Am sorry Manville has not replied promptly to your letter in reference to the Indian Lands. It would seem that he has, in this case at least, carried his habit of not keeping up his correspondence, into a failure to recognize the necessity of prompt action in regard to a business detail. Will call his attention to it with the hope of setting him right on this important matter.*

By the end of 1930, Manville had apparently learned enough to earn the following praise from his father:

Whether you realize it or not I at least realize that you are becoming more and more an essential factor in the management and handling of the business. Already I feel that when either you or your uncle are there our interests are protected in so far as it is possible to protect them and when you are both on the ground I am content, even in a crisis, to be eliminated from the landscape even as a counselor.

A few months before his death in 1933, John Kendrick wrote to his son and gave him possibly the highest compliment he could give:

I note your statement that you "will never get the first idea of business through my head." I think it may interest you to know that as a businessman you suit me exactly and as a son you suit me ideally.

Unlike his father, Manville seemed to have no political aspirations of his own. Nevertheless, he was often touted as one of Wyoming's future leaders. In 1931, after his appearance at the commencement exercises for the Wheatland General Hospital nurses' graduation, Manville was described as:

... a most likeable young man, keenly interested in the great cattle and ranch holdings of the Kendrick family; sharing also his father's love for Wyoming. We were all so glad to meet Mr. Kendrick, Junior, and to have him meet our own young people and many of the younger set of the town, for we feel that it is good for Wyoming that these coming citizens should early begin to share their common interests in the State.

When his father died in 1933, the *Sheridan Press* noted that Manville would most likely "carry on in the footsteps of his father

Manville Kendrick standing near OW chuckwagon, about 1950 (Manville Kendrick Collection)

as one of the northwest's most successful stockmen." It took four years, but after his Uncle Clarence retired in 1937, Manville did indeed take over the reins of the empire, holding them as president of the Kendrick Cattle Company until the ranches were sold in the late 1980s.

Manville was elected to a term as president of the Wyoming Stock Growers Association in 1950, the same group headed by his father in 1912. He was also an officer of the American National Live Stock Association and attended several of their national conventions. In the 1970s, Manville was a member of the Board of

Trustees of the National Cowboy Hall of Fame and Western Heritage Center, into which his father had been inducted in 1958. He also served on the boards of several charitable organizations as well as the Omaha branch of the Federal Reserve Bank.

Manville Kendrick, 1940 (Bachrach, Washington, D.C.; Manville Kendrick Collection)

Manville and Diana were married for fifty-eight years before she passed away in 1987 after a lengthy illness. Manville Kendrick died in September 1992, at the age of ninety-two, at the home he and Diana purchased following their departure from Trail End in 1960. Upon his death, Sheridan residents noted that it was the passing of an era, the tail end of the days of the great western cattleman. Indeed, Manville witnessed many changes during his lifetime: ranching went from horse-drawn to horse-powered, cattle barons were replaced by conglomerates, and increasingly fewer young men were willing to do ranching's hard physical labor in exchange for room, board, and minimum wage.

Through it all, though, Manville tried to maintain the level of gentility with which he had grown up. He once told an interviewer, "There's no aristocracy out here in the real sense. There's a tradition of being a gentleman, and it doesn't matter whether you're a cowboy or a cattleman" (Laton 1984, 134).

Without the driving force of his parents' ambition pushing him ever onward, Manville Kendrick might have remained a cowboy. But even if he had never been a cattleman, he would most certainly have always tried to be a gentleman.

PART 3

❊ ❊ ❊

A DREAM COME TRUE

The Best Trade I Have Made

THE KENDRICK RANCHING EMPIRE

In the late 1800s, thousands of young men arrived in Wyoming and Montana, eager to make their fortunes from the land. Some tried mining or farming while others tried cattle ranching. Because of hard times, hard weather or just plain hard luck, many of these enterprises failed and the adventurers who initiated them disappeared unrecorded into the shadows of time.

Not so John Kendrick. With the right combination of strength, intelligence, timing, drive, and no small degree of luck, he was able to realize his dream of building a vast cattle empire, one to rival those he'd known in his home state of Texas. After spending many years as a cowboy, roundup foreman, and ranch superintendent, John Kendrick finally purchased his own outfit, the OW, in 1897.

The OW was started in the early 1880s by Hiram S. Manville and James S. Peck. Manville, a land developer from Nebraska, had previously owned a freighting company in partnership with Peck, a Milwaukee grain company executive and investor. The OW was originally located near Lusk, Wyoming, in an area occupied by freighter/rancher Oscar C. Wade. According to Manville Kendrick, who was interviewed by David Brinkley in 1961, the origins of the OW Ranch's name are unclear. It might have been named for Oscar Wade, or it might have been named for Old Woman Creek, a small stream running through Wade's land near Lusk.

Rancher Addison Spaugh noted in 1934 that in the early days, Wade's "ranch" consisted of little more than a shack and some cows:

> *There was nothing to the ranch but a squatter's right, as the country had not at that time been surveyed, a dug-out, a small stable, a corral and a small tract of land fenced in for a meadow. But he [Wade] did have 1,200 head of fine cattle. We were not long in making a trade with him.*

In the winter of 1881, based in part on the strength of the OW herds, Manville and Peck joined with several other businessmen to form the Converse Cattle Company, with a capital stock of one million dollars. In addition to Manville and Peck, the founding partners were Cheyenne banker A. H. Converse and cattleman William C. Irvine. Later partners included T. B. Hicks, A. C. May, Samuel McCord, and a Mr. Swallow. All were influential businessmen in their own communities. But like many absentee owners, most—

Below, left: Converse Cattle Company co-founder Hiram S. Manville, about 1900 (Heyn Photography, Omaha; Manville Kendrick Collection)

Below, right: J. S. Peck, secretary/treasurer of the Converse Cattle Company, about 1888 (Armstrong Studio, Milwaukee; John B. Kendrick II Collection)

with the exception of Irvine—did not have a working knowledge of ranching or a clear understanding of range conditions.

By 1884, the OW range was a major center of cattle operations. At the OW roundup that spring there were between twenty and thirty different cattle outfits represented by nearly three hundred men, two thousand saddle horses and over four hundred thousand head of cattle. It took six weeks to round up the herds from the open range and return them to their rightful owners. With those numbers, it is not surprising that the region was quickly overcrowded and overgrazed.

In the winter of 1886–1887, disaster struck in the form of one of the hardest winters the West had ever known. Thousands of cattle died on the Wyoming ranges and hundreds of ranchers, both large and small, were permanently out of business. The Converse Cattle Company was one of these. With its herds gone, its rangeland depleted, and its one million dollars in capital stock nearly exhausted, the partners decided to hire a new superintendent, one who could lead them back from the brink of bankruptcy. After considerable debate, they chose an experienced Texas cowboy named John B. Kendrick.

John B. Kendrick, 1879 (Kirkland Studio, Cheyenne; Hoff Collection)

Kendrick, who had worked the range near Lusk for eight years in the employ of Charles Wulfjen, proved to be an efficient superintendent. He was a hard worker and he demanded the same from the cowboys he supervised. The men were not allowed to drink or gamble while

Kendrick cowboys, 1916 (John B. Kendrick II Collection)

trailing or rounding up the herds; anyone caught doing so was immediately fired. But after the cattle had been delivered to the shipping pens at Lusk, the cowboys were allowed to cut loose:

> *When the cattle were loaded and gone, then the cowboys could go to Lusk for their fun.... [Kendrick] would ride into town and tell the livery stable man that if he saw any OW horses tied in front of a saloon, to take the horses to the barn, feed and care for them and send him the bill. In a couple of days the boys would return to camp and the wagon would pull out for another roundup* (Thorpe, n.p.).

In 1889, just two years after taking over as manager, Kendrick was preparing to move the entire operation to healthier pastures farther north near Sheridan, Wyoming. On March 16, the *Sheridan Enterprise* made note of his arrival:

> *John Kendrick, manager of the Converse Cattle company arrived in town on Sunday. He left the next day ... to look over the Grinnell range on Hanging Woman Creek with a view of*

purchasing same and moving the Converse herd thereto.... This is one of the best ranges in the whole country and we hope satisfactory arrangements will be made whereby the Converse Company will secure it.

The purchase was made and Kendrick moved the OW/Converse operation to an area about fifty miles northeast of Sheridan. During the next few years, he ran the ranch to the best of his abilities, returning whatever profits there were to the absentee owners. In 1897, through the kind auspices of several influential friends, Kendrick finally got his chance to become what he'd always wanted to be: a full-fledged cattle rancher.

In February 1896, some of the Converse Cattle Company's investors unexpectedly sent their bonds in for collection. This could

Kendrick cattle roundup, about 1903 (Sheridan County Historical Society Collection)

have meant disaster for the company; even though the bonds would have to have been redeemed at full face value, they were not worth more than forty cents on the dollar. Rather than go broke, company officials asked the courts to appoint a receiver to handle all company affairs. Will Sturgis of Cheyenne, a well-known cattleman and founding member of the Wyoming Stock Growers Association, was the company's first choice. Through the intervention of influential Nebraska rancher T. B. Hord, however, the Board was persuaded to appoint Kendrick instead. As John wrote to his wife on February 26:

> *When Mr. Hord reached New York the papers were made out asking the court to appoint Will Sturgis of Cheyenne. Mr. Hord talked them into having me appointed in place of Sturgis. So here I am. While I do not as yet know what it will pay me and it may be that I shall wish I were out of it, never-the-less Mr. Hord has once again given substantial evidence of his friendship for me.*

A little less than a year later, on January 28, 1897, John Kendrick wrote a letter to Eula from Chicago, describing how he had finally obtained the company for himself:

> *Dear Sweetheart, … The OW Ranch horses and remnant are ours and here is the way it all came about: We reached Milwaukee last night at half past ten and after an endless delay this morning finally rounded up the Board of Trustees … Mr. Hicks called the meeting to order and proceeded to outline the plan of the sale. When it came to fixing prices we were a good ways apart.… As I named the figures that I would pay for the different properties the excitement was*

tremendous; Swallow and McCord both jumped to their feet and wanted it put on record that they were opposed to a sale on such conditions. I felt almost sure of both Mr. Manville and Mr. Hicks, but needed the third man for a majority. For awhile the outcome was extremely doubtful, but I stood firm and finally won Mr. Peck.

Kendrick was able to purchase all the company's cattle for seventeen dollars a head, calves thrown in for free. Two thousand dollars purchased 213 head of horses, while a 75 percent interest in the ranch and all of its equipment was received for $15,000. When it was all over, Kendrick seemed fairly pleased with himself, as he told Eula:

It is the best trade that I have made since I sold your Papa my only saddle pony to buy me an outfit with nearly 28 years ago. Proportionately that was a better deal than this, but of course there wasn't as much involved.

OW horse cavy, Dutch Creek, Wyoming, 1895 (Manville Kendrick Collection)

Once the final purchase was complete in December 1897, Kendrick moved to expand his holdings in the area, making sure to secure clear title as quickly as possible. He acquired these additional acres through a variety of means. Some were purchased outright from neighbors, for example, while others were obtained through the purchase of soldiers' homestead allotments. Following the Civil War, one of the benefits given to soldiers was the right to file for additional homestead acres on government lands. In partnership with Zachary Hodges of Little Rock, Arkansas, Kendrick arranged to have soldiers (or their survivors) file for homestead claims on government acres near the OW, then sell their claims to Kendrick. Although this practice is often considered questionable at best, it served the purposes of all parties involved. The soldiers, most of whom never intended to move west, got a tidy sum of money while Kendrick obtained title to the land he needed to run his herds. Following the same basic tactics, Kendrick also obtained land through his employees; they would file for homestead acres, Kendrick would make the improvements, and then buy the land from the employees—whether they wanted to sell or not.

Eventually, the various Kendrick ranches encompassed areas of two Wyoming counties (Sheridan and Campbell) and four more in Montana (Powder River, Custer, Big Horn, and Rosebud). They included the Forks, the 76, the E Bar U, Hanging Woman, the 77, Cabin Creek, the Little OW (or Ceded Strip), and Horse Creek. Each ranch or camp had at least a cabin, a barn, and corrals. The larger ones also had guest facilities. Even though each ranch had a full-time manager, John and his general manager Clarence Wulfjen frequently traveled from ranch to ranch in order to check

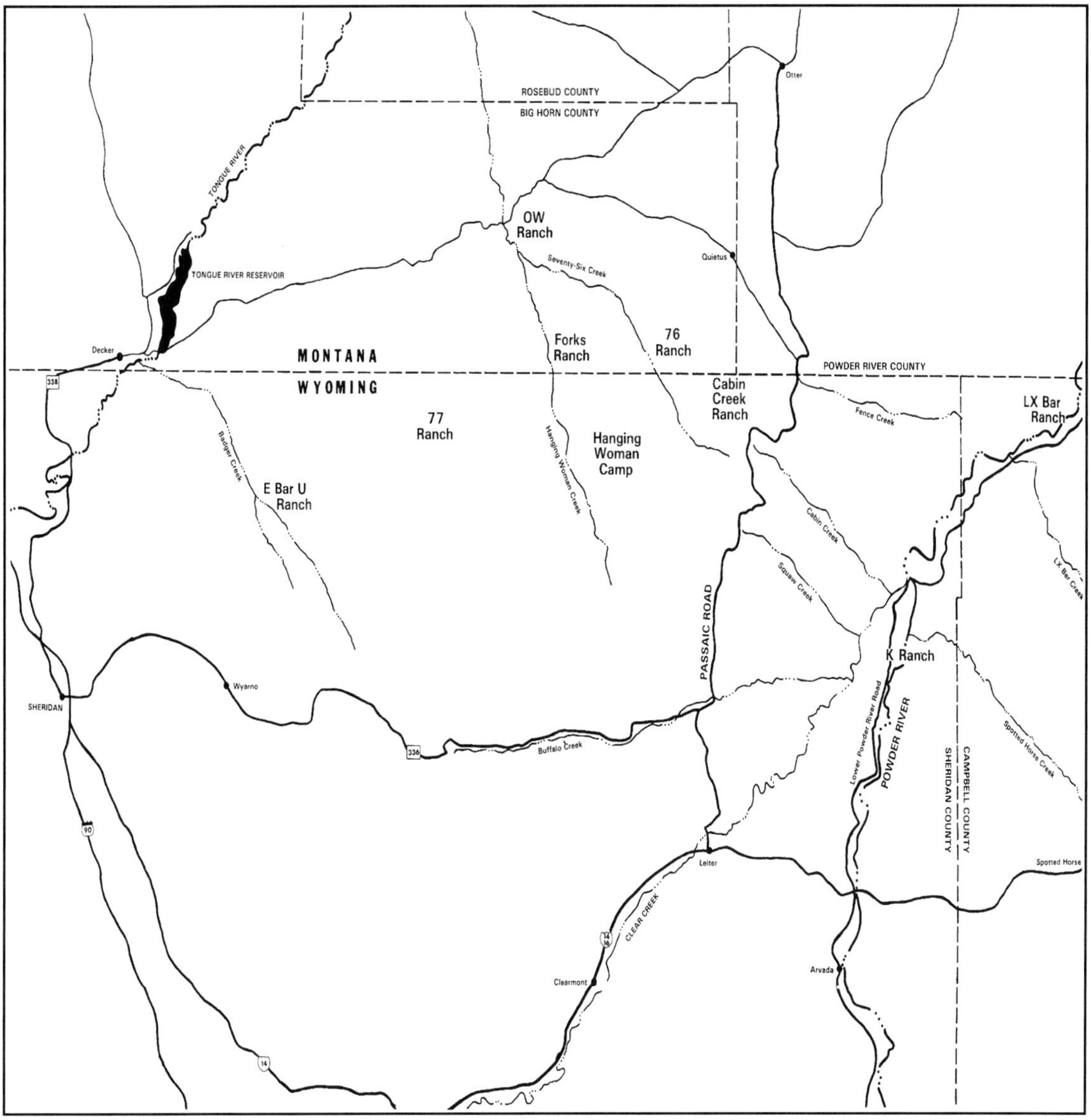

up on operations. Even after the introduction of the automobile, it could take a week to tour all the facilities, and overnight stays were required.

Map of Kendrick ranches in Wyoming and Montana (Val Burgess; Trail End Collection)

When the OW outfit arrived at Hanging Woman, Kendrick found very little in the way of facilities for either his cowboys or his cattle. He quickly began an expansion program and within a few years had a ranch house, bunkhouse, icehouse, horse barn,

Above: OW Ranch, about 1900 (Hoff Collection)

Parlor, OW Ranch, about 1900 (Hoff Collection)

blacksmith shop, and open-shed cow barn, plus circular corrals, outhouses, wells, and gardens. As the "home ranch," the OW was also where John brought his young bride in 1891 and where they raised their children. While Mr. Kendrick spent his time improving the outdoors, Mrs. Kendrick went to work on the main ranch house, furnishing it with almost $900 worth of carpets, curtains, furniture and silverware—an exorbitant sum in those days. Although not a very impressive structure, the main house was jokingly called "The White House" because of the many coats of whitewash that were applied to it over the years.

The second of the three main ranches was the LX Bar, located along Powder River northeast of Sheridan in neighboring Campbell County. Founded in 1878 by the Stanton-Howard Livestock Company, the LX Bar was sold ten years later to A. J. Collins, who in turn sold it to Kendrick in 1902. In 1910, Kendrick hired master stonemason Oscar Husman to construct new buildings including a five-bedroom house, bunkhouse, solar-heated chicken coop, barns, and storage buildings. All were sturdily built with eighteen-inch-thick sandstone walls and two-foot-wide foundations.

Main house, LX Bar Ranch, about 1931 (Tronstad Collection)

Cowboys on the steps of the K Ranch bunkhouse, about 1935 (Lucille Kendrick White; White Collection)

The third main ranch, the K, was the center of the winter range area and principal headquarters for the Kendrick Cattle Company's extensive irrigating operations. Kendrick purchased land containing the K Ranch acreage in 1911 and continued for many years after to acquire adjacent holdings. Located next to Powder River, the K Ranch buildings were also constructed by Oscar Husman, starting in 1924.

The stone buildings at the Ceded Strip (Little OW) near Hardin, Montana, and the E Bar U Ranch northeast of Sheridan were also built by Husman. The other ranches—the 76, 77, Forks, Cabin Creek, Hanging Woman, and Horse Creek—consisted of log buildings and corrals built in a purely utilitarian style. These were the camps where cowboys stayed for months at a time, repairing fences and working with the cattle.

Ranching was, and still is, an extremely difficult business. So much depends on forces outside the ranchers' control. Foremost of these unpredictable forces is weather. No matter what the rancher

Earl Woodbury inside Labrie Cabin at the 76 Camp, about 1920 (White Collection)

does, he can't control the weather and must endure what nature dishes out. Though the disastrous winter of 1886–1887 was one of the worst, hard winters continued to plague Kendrick and other Powder River ranchers in the years to come. As John noted to Texas rancher H. C. Snyder in 1912:

> *Since I have been so long in replying to your very kind favor … you have no doubt concluded that I have been stuck in a snowdrift. In this event you are more than correct for I have been stuck several times within the past few weeks but have managed to wiggle out again.… I have had "one divil" of a time in getting through the winter in this section. In fact … beginning as I did with a large herd of drouth stricken cattle, thin in flesh and greatly reduced vitality, I only wonder that our losses have not been in the nature of a total eclipse.*

When he was not contending with snowdrifts, Kendrick had to worry about the long-term effects of drought. The "drouth" mentioned in his 1912 letter to H. C. Snyder actually began in 1910 and continued for two years. It was a severe drought that caused such overwhelming financial losses that Kendrick endured only by stopping all unnecessary expenditures, including construction at his Sheridan home and at the LX Bar. On July 25, 1911, he wrote to Charles A. Lindner, advising the contractor of his intention to stop work on Trail End:

> *... for the identical reason that compelled me a year ago to delay this construction, I am forced once again to call a halt in all of my business arrangements that compel a demand on my finances. Never in the history of the west have I known a drouth to extend over a period greater than one year's time, and I do not believe that very many white men can recall such a disaster.*

Two decades later, an even bigger disaster struck when the most destructive drought in the history of the United States turned America's heartland into a dustbowl. In the early 1930s, Kendrick's Wyoming and Montana ranchlands dried up like the rest of the country. Even though he had constructed an extensive irrigation system in the 1910s, there was not enough water in the creeks to fill the K Ranch reservoirs. Once drought set in,

Kendrick longhorn cattle, about 1930 (Lucille Kendrick White; White Collection)

grasshoppers by the millions swarmed onto the rangelands, eating everything in sight—or at least anything a cow would be interested in consuming. The Kendrick cattle quickly lost both their weight and their market value.

K Ranch flume, 1914 (John B. Kendrick II Collection)

During the 1930s, ranchers from across the country were experiencing similar ravages and many were simultaneously liquidating their herds. The livestock markets were quickly flooded with underfed cattle and prices plummeted, making returns too small to cover the cost of shipping. As a result, Kendrick and his neighbors slaughtered thousands of cattle and sheep rather than watch them die of hunger and thirst. Although some were skinned, most of the carcasses were simply dumped in the deep breaks cutting through the Powder River landscape.

Another obstacle on the rancher's road to success was the large number of four-footed predators prowling the open range. In early 1912, Kendrick agreed to pay Arvada, Wyoming, trapper A. E. Tompkins "the same bounty that we agreed upon last year, which, if not mistaken, was $10.00 for pups and $20.00 for old wolves." Kendrick paid nearly $1,000 a year towards the extermination of wolves and coyotes from his lands. Al Breckenridge, a trapper based in Otter, Montana, frequently worked for the OW. In 1912, he asked Kendrick to hire him once again:

> *... during my two months of wolfing I have killed 27 pups and four old wolves ... as you have done more to destroy the wolves than any other cow man in the country you [undoubtedly]*

Trapper Charley Florey and collection of wolf and coyote pelts, OW Ranch, 1914 (Hoff Collection)

understand the damage the wolves do. If you feel like donating me a private bounty I will appreciate any bounty that you feel like giving.

Kendrick and his neighbors regularly paid the private bounties. With cattle fetching hundreds of dollars a head and losses to wolves numbering several dozen, Kendrick obviously felt it was worth the investment.

Though willing to confront the wolf at his door to protect his herds and weather nearly any kind of storm or drought for the sake of his ranches, Kendrick was unable to spend the amount of time he would have preferred working with the business. After he became active in politics, he had to depend on his family—including brother-in-law Clarence Wulfjen, nephews Tom Kendrick and Tom Reavis, and son Manville—to take care of the extensive operations. Even from his Cheyenne and Washington offices, however, Kendrick was able to keep an eye on the condition of his empire. Frequent correspondence, such as this letter to Clarence Wulfjen in 1921, indicates that he tried to stay involved in every phase of the ranching operation, from money to manure, with the two being occasionally combined:

> *... the average man, even though he be a superintendent or foreman of a ranch, is not a financier and he doesn't know even a little bit about trying to economize.... One of the impressions I brought away from the K was a man hauling manure from the barn for a mile and a half to stop a wash in the road when every ton of that manure is worth five dollars on the land and when the wash could be stopped with one-tenth of the time with a man, a team and a plow. These sorts of things indicate lost motion that costs all kinds of money.*

John B. Kendrick with brother-in-law Clarence Wulfjen, about 1931 (Tronstad Collection)

As a United States senator, Kendrick worked hard for the benefit of the western rancher. Nevertheless he understood that legislation, no matter how well-intentioned, could not provide every protection the rancher needed:

> *... there is no salvation for the farmer in legislation. Laws are useful and necessary but passing them will not take any farmer out of the rut of depression. There is only one way for him to work himself out and that is by his own individual effort and the sweat of his brow. It seems hard to say it sometimes, but it is true* (Wing 1927, 53).

If anyone knew about the rewards of individual effort, it was John Kendrick. He had triumphed in his own life's work by breaking away from those around him and following his own counsel.

He nevertheless recognized how important cooperative efforts could be in making a large operation such as the Kendrick Cattle Company a success over the long haul. He did not believe in absentee ownership—that was part of what had gotten the Northern Plains ranchers into so much trouble in the 1880s. Nor did he care for the idea of outside involvement. Therefore, in 1923, John entered into a partnership with his wife and children, called the Kendrick Livestock Company (later the Kendrick Cattle Company), wherein:

> *... at all times during the continuance of said partnership [for fifty years and so long as any two of said partners shall survive and desire to continue such partnership], each partner shall give reasonable time, attention and attendance to and use reasonable endeavors in the business of said partnership and shall exert himself for the joint interest, profit and advantage of said partnership.*

Kendrick's children and grandchildren clearly loved the ranches and wanted to spend time there. On August 11, 1914, for example, Rosa-Maye stated in her diary, "Our new house is beautiful, but oh you Ranch! We love you best after all." Later, her children and Manville's spent considerable time riding the ranges and growing into management roles. By sharing the responsibility for the ranches as well as the profits, Kendrick allowed the empire he had built to survive long after his death.

But legally binding contracts do not themselves provide the heart and will needed to keep a ranch running. Though the family benefited for years from its prosperity, the great Kendrick cattle empire eventually crumbled. In late 1988, almost one hundred years after John Kendrick first brought his thirsty herds to the waters of Hanging Woman, family members made the decision

Kendrick grandchildren at OW Ranch, about 1940 (Hoff Collection)

to discontinue the operation. Livestock and equipment were disposed of at public auction, and the Kendrick Cattle Company land, divided into six properties, was sold for an estimated eight million dollars. At the time of sale, the KCC employed twenty to thirty people and ran about four thousand head of cattle on over 209,000 acres. The company was also involved with mineral production, having several oil wells situated on KCC land.

The importance of the ranches to the Kendrick family cannot be denied. It was the prominence that Kendrick gained through his affiliation with the Wyoming Stock Growers Association that allowed him to attain his influential political position in Wyoming. It was the very knowledge that the ranches existed that made it possible for Manville to endure his time away at prep school, Rosa-Maye her time at finishing school, and John his time in Washington. Finally, it was the money raised through the sale of the ranches' cattle that made the construction of Trail End possible.

A Beauty in Design and Finish

BUILDING TRAIL END

In 1895, two years prior to his purchase of the Converse Cattle Company holdings, John Kendrick bought a plot of land from developer A. J. Nielsen. Located on a barren hill in the northwest portion of town, the treeless lot overlooked both the city of Sheridan and the surrounding countryside in two directions. Putting it in his wife's name, Kendrick held on to the undeveloped property for fourteen years. Finally, on September 13, 1908, a short article in the *Sheridan Post* notified the citizens of Sheridan that yet another prosperous area rancher was ready to make his move and build a house in town:

> *John B. Kendrick has accepted the plans for his magnificent new home on [Nielsen] Heights. Teams are now excavating for the foundation, and while the work will take several months, it will be pushed as rapidly as possible.*

Ten months later, in July 1909, local builder Frank Ferguson of Ferguson and Pearson Contractors was asked by a newspaper reporter to discuss new homes being constructed in Sheridan:

> *We are, as you doubtless already know, building for Mr. J. B. Kendrick, a magnificent residence, which was designed by architect McAlister. The house is designed on broad and gen-*

Newspaper advertisement for Nielsen Heights, 1893 (Trail End Collection)

erous lines … and the exterior treatment of the same is attractive, to my mind unusually so, and pleasing to the eye.… The house, which is well situated in the center of a large lot of several acres, will cost, including the stable, about $40,000, and will take nearly a year to complete.

Unfortunately, Ferguson's timeline was not too accurate. Because of construction difficulties, low cattle prices, conflicts between designers, and a string of labor disputes, it would take four more years and four times the expense before the Kendrick family would be able to move into their new home.

The Kendricks spent about $165,000 on their home at a time when local classified ads showed that an average-sized three-bedroom house in downtown Sheridan, with running water, electricity, and a garage, could be had for $3,000. Despite his naturally frugal

East façade, 1913 (Fuller Studio; Sheridan County Historical Society Collection)

nature, John Kendrick did not seem overly concerned with the cost, at least not early on. As he told his interior designer on April 10, 1911:

> *... while inclined to avoid any unnecessary expense, I deem it only fair to you to say at this time that I am not in the least inclined to avoid any outlay that would increase the beauty or practical utility of the house when it is finished.*

That same year, Kendrick again displayed his willingness to invest money in his home when he contracted with Minneapolis landscape architects Anthony Morell and Arthur Nichols for the development of landscape drawings for the property. Even though the cost was far more than he had expected, Kendrick was generally pleased with their plans and, on June 9, gave permission for the work to proceed:

> *... in reference to the terms under which you propose to prepare the scheme for the grounds around my residence ... I will state that while the time required according to your estimate*

East grounds, about 1916 (Hoff Collection)

will make the aggregate cost of this work in excess of what I hoped it would be, I am, nevertheless, going to accept your proposition, and in doing so I propose to rely upon your integrity in seeing that the total expense is not greater than is justified by the work to be done.

The primary intent of Morell and Nichols' plan—the first known landscape architecture plans prepared for any private residence in the state of Wyoming—was to create an informal outdoor space that would serve as a natural setting for the home itself. The pair eschewed formal gardens and intricate hedge arrangements in favor of informal groupings of native trees with paths meandering throughout:

The arrangement of groups of trees and shrubs aims principally in giving setting to your residence and other buildings as well as to frame the grounds and give you as much privacy as possible and to show up the lawn to the best of advantage thus forming an harmonious whole.... if there is anything which you would like to change please mark on the plan and return same to us so that we could get a clear idea of your wishes (Morell and Nichols 1911).

Getting a clear idea of John Kendrick's wishes was the goal of every contractor involved in the Trail End project. Though Eula Kendrick was actively involved in its design and construction, Trail End was apparently John Kendrick's idea from start to finish. Mary Kendrick Morgan, a relative from Texas who lived on the family ranches during the planning phase of the project, later told Manville that Eula was not necessarily eager to take on the responsibility of such a large home:

I helped your mother a little on the plans when I was with you folks and she said then the house was going to be a big responsibility. I think that your Dear Father wanted the big house much more than she did. I heard your grandmother [Ida Peeler Wulfjen] tell your mother not to oppose him about the house, that he had worked hard and building that house had been a dream of his for a long time (Morgan 1960).

Lyman Brewster, member of a neighboring ranch family with close ties to the Kendricks, also believed that Eula was not the driving force behind the house. He recalled, "Aunt Eula never liked it, and would have preferred the ranch, in Montana, of course" (Brewster 1982). Manville concurred, noting in a 1987 interview that his mother "didn't take to it too strongly" while it was being built.

Trail End was designed by Billings, Montana, architect Glenn Charles McAlister. Almost completely self-taught, McAlister was chosen from a pool of professionals who had submitted their drawings to the Kendricks as early as 1907. He had adequate credentials, having already designed and built two of Sheridan's more impressive structures: the Sheridan County Courthouse and a large private residence known as "Mount View."

Architect Glenn McAlister, about 1955 (Alfred and Fabris; Sheridan County Historical Society Collection)

McAlister's design called for the front of the house to be symmetrical with a formal entrance reminiscent of large estates in Great Britain, New York, and Virginia. The sides and back were given the more random arrangement of porches, balconies, windows, and chimneys typical of late-nineteenth-century Victorian houses.

Trail End is an imposing structure with thirty rooms encompassing over thirteen thousand square feet. There are three floors plus a basement, attic, four balconies, and four porches. Building materials included beige Roman-style brick from Kansas (held together with specially tinted red mortar), white Indiana limestone, and red Missouri roofing tiles. The house sits atop a foundation made of Montana granite.

The home was built in the Flemish Revival style, recognizable by the presence of curvilinear gables most often seen in the border areas between France and Belgium. It was an unusual choice, clearly made to impress the community. Certainly the cosmopolitan nature of the new residence was not lost on the citizens of Sheridan or reporters at the *Sheridan Post.* As one noted in November of 1909:

> *Numerous other buildings are in the course of construction, including a number of dwelling houses, indicating that Sheridan*

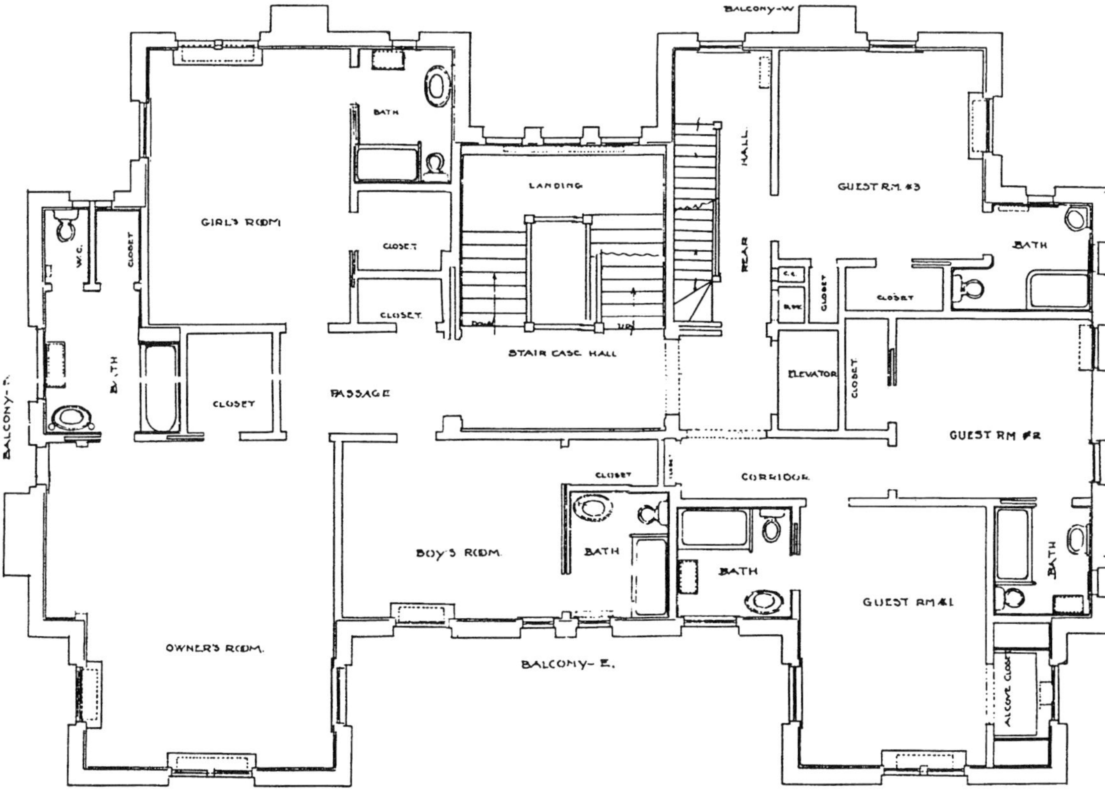

Second floor plans, 1911 (Trail End Collection)

is and will continue to be a city of homes. Among these is the magnificent residence of J. B. Kendrick on [Nielsen] Heights, which would adorn Fifth Avenue in New York or the lake front of Chicago.... The new home is a beauty in design and finish, being a credit to the enterprising businessman who built it, and the city as well.

Just as the home's exterior reflects diverse stylistic influences, the interior also exhibits a variety of design influences. Victorian, Edwardian, Prairie Style, and Arts and Crafts elements can be seen throughout. Even with this eclectic design, there was to be a certain air of simplicity about the house, a withdrawal from the cluttered, fussy look of the 1890s. Although difficult to tell from the intricate woodwork and furnishings in evidence at Trail End,

Kitchen, 1913 (Fuller Studio; Trail End Collection)

simplicity really was the goal, as stated by John Kendrick in an April 1911 letter to his interior designer:

> *Concerning my wishes as to the general plan of the interior work, I should like to have it treated along the lines of the utmost simplicity consistent with the best artistic effect, in a character that would not only be livable when we first moved into the house but would continue to grow in favor as we become more and more accustomed to it.*

To create their "livable" home, John and Eula relied on manufacturers and designers from all over America. Though contractors Frank Ferguson and N. A. Pearson were in charge of the local work force, expert craftsmen had to be called in for some of the specialty tasks. Like most of the designers, decorators, and fabricators, these workmen were based well over five hundred miles away. Since the Kendricks acted as their own general contractors (to take advantage of contractors' discounts), it was up to them to coordinate all activities—a monumental feat they found difficult to accomplish alone.

Visiting the individual manufacturers was costly and time-consuming. Therefore, the Kendricks used catalogs, diagrams, drawings, and samples to make many of their decisions. Occasionally, though, John or Eula would have to meet with the vendors face-to-face, traveling by train to Omaha, Grand Rapids, and Chicago on a regular basis.

John and Eula were well aware of the fact that they did not know everything it took to put together a spectacular house. In 1911, they enlisted the aid of an interior designer to finalize the decorating plans and provide guidance on the overall look of the house. D. Everett Waid, a New Yorker who later served as head of the American Institute of Architects, was hired to lead the Kendricks toward the tasteful. This he was not shy to do. In May 1912, Waid voiced his concerns about the fireplace finishes:

> *Regarding the fireplaces, I would say that to my taste, both for aesthetic and practical reasons, onyx would be very objectionable. A very quiet, dignified and yet rich effect can be obtained appropriate to the style of the interior design by a proper selection of either tile or marble.*

Opposite—

Top: Dining room, 1913 (Fuller Studio; Harmon Collection)

Bottom: Drawing room, 1913 (Fuller Studio; Trail End Collection)

The Kendricks usually followed Waid's suggestions (they chose a dignified Italian marble for the fireplaces mentioned above), but they were not always sure they wanted to. In early April 1911, Eula wrote to Waid about his suggestions for the living room:

If I am not mistaken, you suggested paneling that room also when we were talking it over together. I did not think then that Mr. Kendrick would care for it, but he seems to think he would like it better since going over these drawings, but we have decided to leave it to your good taste....

Even though Kendrick now had two architects working for him, the men were not always available. Waid was located on the East Coast, and McAlister was rarely in Sheridan, preferring to spend time on his ranch in Montana. This was quite vexing to John Kendrick, who noted in a July 17, 1911 letter to manufacturer Charles A. Lindner:

I will say that I have had no end of trouble in trying to worry through this work under the direction of McAlister, and since we have arranged for the assistance of the New York man, he is so far away that I do not get very much better results from him. At the present time McAlister is out on a ranch some place....

In the end, it fell to Lindner to coordinate efforts between the woodworkers, the furniture makers, and the interior finishers. Lindner, president of the Lindner Interior Manufacturing Company of Grand Rapids, Michigan, was hired in late 1909 and quickly

Lindner letterhead, 1911 (Trail End Collection)

gained the respect of the Kendrick family. In a May 1911 letter to Waid, John referred to Lindner as "one of the most satisfactory men with whom I have dealt." He and Eula had the utmost confidence in Lindner's abilities, as evidenced in this December 1909 letter to the contractor:

> *The question of our interior woodwork has given both Mrs. Kendrick and myself an endless amount of anxiety, but since our talk with you we have a feeling of complete assurance as to the outcome so that we have practically dismissed it from our minds.*

Under Lindner's personal supervision, his firm not only manufactured and installed all the woodwork in the house, but also took charge of matching the furniture finishes to the walls, locating a stained-glass firm, and coordinating final installation with the interior decorators. Lindner's job was somewhat simplified by the fact that several of the manufacturers with whom he dealt were located in and around Grand Rapids.

While Lindner worked on putting the house together, McAlister and Waid seemed to spend their time sniping at one another. In letters to the Kendricks, Waid was openly scornful of McAlister's talents and ideas, probably because of the Montanan's lack of formal training. In the end, both architects faded from the scene and Lindner became the driving force behind the home's completion.

Lindner and Kendrick worked together to incorporate many features not found in the typical Wyoming home. John Kendrick was a progressive man and he wanted a progressively modern home. If a piece of equipment or a new invention made a task easier or less expensive, he wanted to make use of it. He had done so at the ranches for many years, so it was not surprising that he

Above: Foyer stairs, 1913 (Fuller Studio; Trail End Collection)

Master bath, 1913 (Fuller Studio; Sheridan County Historical Society Collection)

took advantage of many of the new technologies available to homeowners in the early 1900s. Beyond the basic luxuries of electricity and hot and cold running water, the new house contained a variety of modern, labor-saving devices including an intercom system, stationary vacuum cleaner, dumbwaiter, laundry chute, annunciator, and telephones.

Although Trail End took over five years to build, work was not in progress that entire time. The first major work stoppage came in August 1909 when a delay in the delivery of foundation granite caused John Kendrick to advise Glenn McAlister:

> *The thing that is worrying us, however, is the delay in the arrival on that granite. We do not hear a single word from it nor from the firm contracting to deliver it. Unless it reaches us in the meantime every man on the building will be paid off and the work entirely stopped on Saturday night. I cannot conceive of a more exasperating situation than this. We have the workmen engaged, the weather is fine and our time in which we can safely build is running short.*

In 1910, work was halted for nearly two years when a drought caused such serious reversals in the cattle market that Kendrick had no extra cash to spend. As he noted to Charles Lindner on July 25, 1911:

> *The reverses met in my business within the past eighteen to twenty months have been real and not theoretical and I feel of course that it is best to bend all my energies and employ all the available funds at my command in protecting the business on which I depend rather than assuming further obligations that would be at this time in the nature of an extravagance.*

Trail End under construction, 1909 (Murff Collection)

In addition to curbing his "extravagances," John Kendrick also felt the need to cut back on such basic expenditures as living expenses. When the Kendricks first moved into Sheridan in 1908, they rented rooms at the Sheridan Inn and later rented a small house. Since they expected construction to take only a year or so, this did not seem like such a hardship. As time stretched on, however, the Kendricks realized they needed someplace else to live. Since it was already built and paid for, Eula got the idea to move into the carriage house:

> *The planning, erection, and completion of this house took longer than had been foreseen—not an unusual experience in home building, but one which proved trying to a woman who disliked procrastination in any form as much as Eula Kendrick. She decided to go and live in the barn, which was nearly completed but which had not yet been put to its intended use, and oversee the work* (Keyes 1931, 51–52).

Carriage house, 1913 (Fuller Studio; Trail End Collection)

Finished in the same Flemish style as the main house, the Kendricks' carriage house was easily adapted for the family's use. The horse stalls, complete with polished wood floors, concrete water troughs, cast-iron hay chutes, and elaborate railings, served as bedrooms for Rosa-Maye and the cook. Another was put to use as a temporary kitchen. John and Eula Kendrick slept upstairs in the hayloft while Manville camped out on a cot in the main carriage room. When company came to town, the semi-detached chicken coop, never used for its original purpose, was pressed into service as a guest bedroom.

The family did not seem embarrassed by their circumstances; rather, the opposite seemed true. In March 1912, John wrote to his former business partner, A. S. Burrows, unabashedly inviting him to stay in their "barn" on his next visit:

The Kendricks are doing some tall figuring on getting their house done by the coming of summer but it is not material that

they should do so. It would serve you exactly right to have to live in the barn for a short time where we have lived so long, when you come to see us.

Construction resumed in the spring of 1912 and hopes ran high that the family would soon be able to move into their new home. Unfortunately, 1912 brought even more problems. Along with a misplaced railcar full of woodwork (eventually located on a siding in rural Michigan), Trail End's principal contractor was plagued by labor disputes. As Charles Lindner noted in late November 1912:

We have never had a Year when so many complications arose, especially pertaining to the question of labor, and we have had almost a continuous round of labor difficulties here and in Chicago the last year. It appears now that the settlement of these conditions in Chicago will soon be reached satisfactorily all around, but the demands and conditions of the Unions now-a-days are such that it means you are hardly able to operate your own business the way you know it should be done.

Nevertheless, Lindner remained optimistic that the Kendrick family would soon have a finished home. He wrote optimistically on November 22:

We note what you say about the completion of your home so that you might be settled by the first of April [1913], and wish to state, that so far as we can see at the present time, and unless something unforeseen happens, there should be no reason why this should not be accomplished.

Opposite—

Top: Interior of carriage house, 1910 (John B. Kendrick II Collection)

Bottom: Master bedroom, 1913 (Fuller Studio; Trail End Collection)

First landscaping, 1914 (Fuller Studio; Sheridan County Historical Society Collection)

It took a little bit longer, but finally, on July 28, 1913, Rosa-Maye Kendrick's sixteenth birthday, the family moved the last of their things into Trail End. Though work continued for awhile—the furnace was not operating correctly and the landscaping was not finished—at least the family was finally at home. Not surprisingly, the Kendricks found Trail End to be quite intimidating, particularly after the relative simplicity of both the OW Ranch and the carriage house. The grandeur of the mansion presented an incredible change for the entire family, especially sixteen-year-old Rosa-Maye Kendrick, who noted in her diary on July 30: "Have been in town two or three days now. House was bewildering when I first came in. Am just beginning to feel at home last day or so."

A series of teas and open houses soon followed during which friends and family members were invited to see what five years of hard work had created. The home's formal open house was held on New Year's Day, 1914. The next day, the *Sheridan Post* described the scene:

Winter at Trail End, about 1920 (Elsa Spear; Bilyeu Collection)

"Trail End," the beautiful new home of Mr. and Mrs. J. B. Kendrick on Nielsen Heights, was formally opened to their many friends between the hours of three and six on New Year's Day. During this time about two hundred guests called and enjoyed the hospitality of this elegant home.

These first visitors to Trail End found an interior grand enough to match the impressive exterior. The first floor contained spacious rooms with twelve-foot-high, hand-painted ceilings, large plate-glass windows, and solid oak floors. Custom-made brass and silver chandeliers and wall lights added a modern glow to the dark, exotic hardwoods lining the walls. Fine furnishings and accessories were placed decoratively throughout the house, along with hothouse flowers and potted greenery:

In the spacious drawing room, which was adorned with American Beauty roses and potted ferns, Mr. and Mrs. Kendrick received their friends.... In the library adjoining

Library, 1913 (Fuller Studio; Trail End Collection)

the drawing room Mr. Clarence Wulfjen assisted the hosts. Festoons of pink and white carnations and smilax prevailed in the artistic dining room where dainty refreshments were served. Potted plants and cut flowers were used as general decorations in the halls.

Guests were bombarded with color in every room. Red, rose, green, gold, and blue velvet and damask draperies hung at the windows and doors of the foyer, drawing room, dining room, and library. Blue hand-painted friezes decorated the dining room walls while pink French silk damask lined the walls of the draw-

Ballroom, 1913 (Fuller Studio; Sheridan County Historical Society Collection)

ing room. Ruby Wilton–style carpeting covered the foyer floor and extended up the wide staircase to the second and third floors, providing a rich color contrast to the green, blue, and brown oak leaf–patterned paper lining the walls above the dark oak wainscoting.

The upper floors contained three bedrooms for the family, three more for guests, and three for staff. Each of the guest and family rooms had private baths while the staff shared a single bath on the third floor. The top floor also contained a ballroom with an open-beam vaulted ceiling, brick fireplace, and musicians' loft:

> *A pleasing feature of the reception was the dancing in the ball room on the third floor, where Christmas greens were the predominating decoration. Here Miss Rosa-Maye Kendrick presided.... Master Manville Kendrick and Master Francis Williams presented each guest with an appropriate favor of the occasion in the form of a booklet containing New Year's greetings from the Kendricks and a picture of "Trail End."*

Trail End as seen from Lewis Street, 1927 (Hoff Collection)

From the time the Kendricks moved into their house, it was known as Trail End. While there is little documentary evidence to prove it, family legend states that John Kendrick named the house Trail End because it was to be the *end* of his personal *trail*, the one that led a penniless orphan from the shores of Matagorda Bay to the halls of a grand mansion. Author and family friend Frances Parkinson Keyes thought the name was symbolic of the family's success:

> *At last the new house was finished, a beautiful house, spacious and sunny, furnished with exquisite taste and restrained elegance. The Kendricks had built for all time and built well; and they named the home "Trail End," not only because it really marked the end of the old cattle trail, but also because they thought that it marked the summit of their own ascending fortunes* (Keyes 1931, 52).

Others who knew Kendrick and his story certainly thought the name was appropriate. As businessman Wilbur Burgess stated to Eula Kendrick in early 1914:

I think the name Mr. Kendrick has chosen "Trail End" is certainly very appropriate and original and so different from what most people would select. I sincerely hope that the trail may not end for a great many years to come.

John B. Kendrick at Trail End, 1931 (John B. Kendrick II Collection)

Little did Kendrick know that Trail End was just the beginning. Barely eighteen months after moving into their new house, the family found itself living in the Wyoming Governors' Mansion in Cheyenne following John's victory in the 1914 gubernatorial election. Two years later, Kendrick was chosen to serve in the United States Senate and the family moved to Washington, D.C. After that, Trail End became a vacation home for John and Eula, one they longed for when the affairs of state became burdensome.

John Kendrick died before he could return to Trail End permanently. In one way, however, the home did become the end of his trail: following his death in 1933, a private family funeral service was held for John Kendrick in the drawing room of his dream house. As the editor of the *Sheridan Morning News* noted:

It was "meet and just" that the sod, that formed the foundation for Kendrick footsteps to trod to fame and honor ... should eventually enshroud him in his final sleep at the threshold, so to speak, of his own "Trail's End."

EPILOGUE

❋ ❋ ❋

KEEPING THE DREAM ALIVE

Perhaps a Museum

SAVING TRAIL END

Though the Fates conspired to keep him from living at Trail End for more than a few months at a time, John Kendrick was hopeful that his family, at least, would be able to enjoy his dream home. In a successful effort to convince his married son to live at Trail End instead of moving out and building a new home, John Kendrick wrote to Manville in May 1932:

> *... the house was not built primarily for my own comfort and enjoyment, but in the hope that succeeding generations would have the privilege of living in it and enjoying its benefits and advantages.... When the time comes that my own descendants do not care to use and keep the home I hope they will find it possible to deliver it into the keeping of the City of Sheridan, perhaps as a museum or art gallery....*

Only two succeeding generations had the privilege of living in the home. By 1968, Trail End (or the "Kendrick Mansion" as it is known throughout Sheridan) had stood empty for seven years. Mrs. Kendrick had died in 1961 and her family had moved on to smaller homes of their own. Because of regular housekeeping, there were no layers of dust on the furnishings; but there was a layer of apathy concerning the building's future.

The City of Sheridan briefly considered buying the Kendrick

Three generations of Kendrick men: John B., Manville, and John B. II, Trail End, 1932 (Manville Kendrick Collection)

Mansion and running it as an art center. When that plan did not work out for financial reasons, the local school district contemplated purchasing the three-story brick structure for use as an educational facility. That effort also ended because of financial concerns. No one seemed to have either the time or the money to turn such dreams into reality. Finally, it looked like the building would be torn down to make room for condominium housing. A public auction was held and the house was stripped of its contents, including furnishings, light fixtures, and rugs.

Fortunately for Sheridan's "Castle on the Hill," a rescue team arrived in the form of the Sheridan County Historical Society. With money donated and lent by both private individuals and government agencies, the Society purchased the land from the Kendrick heirs who in turn donated the structure and what contents remained. With the efforts of an all-volunteer staff, Trail End was turned into a community museum housing original furnishings, archaeological collections, local memorabilia, and fine works of art.

Opposite—

Top: Rosa-Maye's bedroom, 1913 (Fuller Studio; Trail End Collection)

Bottom: Trail End, about 1945 (Hoff Collection)

Eventually the rising costs of utilities and repairs became too much of a burden for the Society's all-volunteer staff. Changes had to be made or the building they had fought so hard to save would be lost once again. In 1982, after months of internal debate, the Society voted to give up Trail End and turn the keys over to the Wyoming State Archives, Museums and Historical Department (now the State Parks and Historic Sites Division of the Wyoming Department of State Parks and Cultural Resources).

Shortly after the state took over operations, the decision was

made to remove the museum exhibits and return the structure to its historic appearance. Trail End had suffered very few alterations since its completion in 1913, thus making it an ideal candidate for a historic house museum. Built by a cattle baron for a cattle baron's family, the fine workmanship, opulent style, and close attention to detail showed the pride the Kendrick family had in their home.

Because of the generosity of the extended Kendrick family and the Sheridan community, many original furnishings were returned to Trail End, thus affording visitors a remarkably true vision of what once was just one cowboy's dream.

Bibliography

BOOKS:

Atherton, Lewis. *The Cattle Kings.* Lincoln: University of Nebraska Press, 1961.

Boyd, David. *The History of Greeley and the Union Colony.* Greeley, Colorado: 1890.

Brown, Mabel, ed. *First Ladies of Wyoming, 1869–1990.* Cheyenne: Wyoming Commission for Women, 1990.

Burroughs, John Rolfe. *Guardian of the Grasslands: The First Hundred Years of the Wyoming Stock Growers Association.* Cheyenne: Pioneer Printing, 1971.

Garst, Shannon and Warren Garst. *Cowboys and Cattle Trails.* Chicago: Wheeler Publishers, 1948.

Harmon, Rosa-Maye Kendrick. *Intimate Letters from London.* Denver: Mrs. John B. Kendrick, 1928.

King, Bucky. *The Empire Builders: The Development of the Kendrick Cattle Company.* Sheridan, Wyoming: By the Author, 1992.

Larson, T. A. *A History of Wyoming.* Lincoln: University of Nebraska Press, 1965.

Memorial Services Held in the House of Representatives of the United States, Together with Remarks Presented in Eulogy of John B. Kendrick, Last a Senator from Wyoming. Washington: Government Printing Office, 1934.

Rollinson, John K. *Wyoming Cattle Trails.* Caldwell, Idaho: Caxton Printers, 1948.

Stout, Tom. *Montana: Its Story and Biography.* Chicago: American Historical Society, 1921.

PRIMARY SOURCES:

Harmon Collection. Trail End State Historic Site, Sheridan, Wyoming.

Hoff Collection. Trail End State Historic Site, Sheridan, Wyoming.

Kendrick Cattle Company Collection. Wyoming State Archives, Cheyenne.

Kendrick Collection. American Heritage Center, University of Wyoming, Laramie.

Kendrick Collection. Wyoming State Archives, Cheyenne.

Kendrick Family Genealogical Files. Trail End State Historic Site, Sheridan, Wyoming.

Kendrick, Manville. Interview by David Brinkley, OW Ranch, Montana, 1961. Harmon Collection, Trail End State Historic Site.

Kendrick, Manville. Interview by Margot Liberty, Trail End, Sheridan, Wyoming, 15 July 1987. Kendrick Family Research Files, Trail End State Historic Site.

Kendrick, Rosa-Maye. Diaries, 1913–1926. Eula Harmon Hoff Collection, Trail End State Historic Site, Sheridan, Wyoming.

Thorpe Collection. American Heritage Center, University of Wyoming, Laramie.

Trail End Construction Documents. Wyoming State Museum, Cheyenne.

White Photograph Collection. Trail End State Historic Site, Sheridan, Wyoming.

PERIODICALS:

Carroll, Eugene T. "John B. Kendrick, Cowpoke to Senator, 1879–1917," *Annals of Wyoming*, spring 1982.

Cutting, Malcolm C. "The Northwest Trail to the Senate," *The Country Gentleman*, March 1926.

Gilbert, Clinton W. "The Sagebrush Senator," *Collier's*, 30 June 1928.

Henderson, Harry B., Sr. "Governors of the State of Wyoming: John B. Kendrick," *Annals of Wyoming*, 1943.

Kendrick, John B. "The Unromantic West," *The Country Gentleman*, 1924.

Keyes, Frances Parkinson. "Over the Trail to the Senate," *Delineator Magazine*, April 1931.

Knappen, Theodore M. "The West at Washington: Significant News from the Nation's Capital," *Sunset Magazine*, June 1924.

Laton, Michael. "Ah! Wyoming," *Town & Country*, July 1984.

Painter, Fannie. "From Log Cabin to an Executive Mansion," *National Magazine*, January 1928.

"Senator Kendrick's Visit," *Bulletin of the Wheatland General Hospital*, May 1931.

Wing, Andrew S. "John B. Kendrick, The Orphan Who Took the Texas Trail to a 200,000 Acre Ranch and the U.S. Senate," *Farm and Fireside*, November 1927.

NEWSPAPERS:

"A Brilliant Wedding," *Greeley (Colorado) Tribune*, 21 January 1891.

"All Washington Is Tense with War Activity," *Cheyenne Daily Leader*, 12 May 1917.

"Another Country Heard From," *Sheridan Post*, 8 July 1922.

"Contractors Are Kept Busy," *Sheridan Post*, 13 July 1909.

Deming, William C. "Stirring Incidents in Life of U.S. Senator Kendrick," *Wyoming State Tribune*, 28 August 1932.

Edwards, Fraser. "Kendrick of Wyoming, Senate's Cattle King," *Washington Herald*, 2 September 1929.

"Endorsed Kendrick," *Douglas (Wyoming) Enterprise*, 18 July 1922.

"Even Old Livestock Records Stand as Tribute to Kendrick," *Sheridan Press*, 7 November 1933.

"Five Years Ago," *Sheridan Press*, 8 September 1913.

Golden (Colorado) Republican, 8 July 1922.

"Good Outlook in Sheridan," *Sheridan Post*, 26 November 1909.

Kendrick, John B. "The Texas Trail," *Wyoming Tribune*, 16 September 1916.

Loucks, John D. "Founder of Sheridan Tells How He Laid Out Beautiful City," *Sheridan Enterprise*, 12 May 1914.

"Miss Kendrick Will Live in England," *Cheyenne Tribune*, 23 January 1927.

Morrow, James B. "John B. Kendrick, Cow-Puncher and Statesman," unidentified newspaper, 14 June 1917. Kendrick Family Research Files, Trail End State Historic Site.

"New Banking Firm," *Sheridan Post*, 16 October 1890.

"Receptions," *Sheridan Post*, 2 January 1914.

"Republican-Boomerang Contributors to Travel," *Laramie Republican-Boomerang*, 26 May 1926.

Saratoga (Wyoming) Sun, 12 November 1916.

"Senator Kendrick Held in High Esteem over Entire Wyoming," *Sheridan Press*, 5 November 1933.

Sheridan Enterprise, 16 March 1889.

Sheridan Morning News, 4 November 1933.

Spaugh, Addison A. "The Famous OW Roundup of 1884," *Lusk (Wyoming) Free Lance*, 14 May 1934.

Stewart, Maud B. "Society," *Sheridan Enterprise*, 7 July 1918.

"Sundance Christened by Miss Kendrick," unidentified newspaper, June 1920. Kendrick Family Research Files, Trail End State Historic Site.

"To Say Farewell," *Sheridan Press*, 8 June 1961.

"West Mourns Kendrick," *Sheridan Press*, 5 November 1933.

"What 'The Commoner' Says of Gov. John B. Kendrick," *Cheyenne State Leader*, 24 October 1916.

"Wife of Native Texan Who Started Teapot Dome Probe Holds Popular Place in Society of Washington," *Austin American*, 17 February 1923.

Wiggins, J. Vinson. *The Cherokee County (Texas) Banner*, n.d.

UNPUBLISHED WORKS:

Barnes, J. Youge. "Constructive Analysis of Miss Rosamaye Kendrick." Unpublished manuscript, January 1923. Kendrick Family Research Files, Trail End State Historic Site.

Carroll, Eugene. "John Benjamin Kendrick, Western Senator." Master's Thesis, University of Wyoming, Laramie, 1977.

Fley, Joann. "John B. Kendrick's Career in the United States Senate." Master's Thesis, University of Wyoming, Laramie, 1953.

Kendrick, Rosa-Maye. "Fragments of My Trip with Daddy." Unpublished manuscript, September 1926. Kendrick Family Research Files, Trail End State Historic Site.

Lewellyn, Michael M. "John Kendrick and the Revival of the Democratic Party in Wyoming, 1910–1914." Master's Thesis, University of Wyoming, Laramie, 1975.

Murray, Robert A. "Management and Interpretive Plan for the Trail End Project." Prepared for the Sheridan County Historical Society, Sheridan, Wyoming, 1972.

Wyoming State Parks and Historic Sites. "Trail End State Historic Site Master Plan." Edited by John Mahoney and Tom Lindmier. Cheyenne, Wyoming: Department of Commerce, 1993.

CORRESPONDENCE:

Breckenridge, Al. Letter to John B. Kendrick, 7 June 1912. Correspondence Files, Kendrick Cattle Company Collection, Wyoming State Archives, Cheyenne.

Brewster, Lyman. Letter to Margot Liberty, 25 August 1982. Kendrick Family Research Files, Trail End State Historic Site, Sheridan, Wyoming.

Burgess, Wilbur. Letter to Eula Wulfjen Kendrick, 10 January 1914. Trail End Construction Documents, Wyoming State Museum, Cheyenne.

Ely, Elizabeth D. Letter to Eula Wulfjen Kendrick, 21 January 1915. Eula Harmon Hoff Collection, Kendrick Family Research Files, Trail End State Historic Site, Sheridan, Wyoming.

Kendrick, Eula Wulfjen. Letter to Manville Kendrick, 2 May 1928. Correspondence Files, Manville Kendrick Collection, Trail End State Historic Site, Sheridan, Wyoming.

_____. Letter to D. Everett Waid, 10 April 1911. Trail End Construction Documents, Wyoming State Museum, Cheyenne.

Kendrick, John B. Letter to Jack (?), 2 January 1914. Correspondence Files, Kendrick Collection, American Heritage Center, University of Wyoming, Laramie.

_____. Letter to Tom Bell, 16 January 1932. Correspondence Files, Kendrick Collection, American Heritage Center, University of Wyoming, Laramie.

_____. Letter to A. S. Burrows, 8 March 1912. Kendrick Family Research Files, Trail End State Historic Site, Sheridan, Wyoming.

_____. Letter to Democratic State Committee, 17 September 1914. Correspondence Files, Kendrick Collection, American Heritage Center, University of Wyoming, Laramie.

_____. Letters to Eula Wulfjen Kendrick, 21 March 1891, 16 April 1891, 11 May 1891, 26 February 1896, 28 January 1897. Correspondence Files, Manville Kendrick Collection, Trail End State Historic Site, Sheridan, Wyoming.

_____. Letter to Josephine Kendrick Woodrel, 1917. Kendrick Family Research Files, Trail End State Historic Site, Sheridan, Wyoming.

_____. Letters to Manville Kendrick, 1 October 1918, 5 October 1918, 19 October 1919, 24 November 1930, 24 May 1931, 24 May 1932, 4 May 1933. Correspondence Files, Manville Kendrick Collection, Trail End State Historic Site, Sheridan, Wyoming.

_____. Letter to Rosa-Maye Kendrick, 10 October 1915. Kendrick Family Research Files, Trail End State Historic Site, Sheridan, Wyoming.

_____. Letters to Charles A. Lindner, 28 December 1909, 17 July 1911, 25 July 1911. Trail End Construction Documents, Wyoming State Museum, Cheyenne.

_____. Letter to H. A. Loucks, 29 September 1919. Correspondence Files, Kendrick Collection, American Heritage Center, University of Wyoming, Laramie.

_____. Letter to Glenn Charles McAlister, 5 August 1909. Trail End Construction Documents, Wyoming State Museum, Cheyenne.

. Letter to Morell and Nichols, 9 June 1911. Trail End Construction Documents, Wyoming State Museum, Cheyenne.

_____. Letter to Helen [illegible]hinson, 25 February 1932. Correspondence Files, Kendrick Collection, American Heritage Center, University of Wyoming, Laramie.

_____. Letter to H. C. Snyder, 22 March 1912. Correspondence Files, Kendrick Collection, American Heritage Center, University of Wyoming, Laramie.

_____. Letter to A. E. Tompkins, 15 February 1912. Correspondence Files, Kendrick Collection, American Heritage Center, University of Wyoming, Laramie.

_____. Letters to D. E. Waid, 10 April 1911, 26 May 1911. Trail End Construction Documents, Wyoming State Museum, Cheyenne.

_____. Letters to Clarence Wulfjen, 18 May 1921, 5 July 1922, 15 December 1922. Correspondence Files, Kendrick Cattle Company Collection, Wyoming State Archives, Cheyenne.

Kendrick, Manville. Letter to Annabelle Hoblit, 17 February 1971. Kendrick Family Research Files, Trail End State Historic Site, Sheridan, Wyoming.

_____. Letter to John B. Kendrick, 25 September 1918. Correspondence Files, Manville Kendrick Collection, Trail End State Historic Site, Sheridan, Wyoming.

Kendrick, Rosa-Maye. Letter to John B. Kendrick, 19 March 1915. Eula Harmon Hoff Collection, Kendrick Family Research Files, Trail End State Historic Site, Sheridan, Wyoming.

_____. Letter to John B. Kendrick, 14 June 1920. Correspondence Files, Manville Kendrick Collection, Trail End State Historic Site, Sheridan, Wyoming.

Lindner, Charles A. Letters to John B. Kendrick, 22 November 1912, 25 November 1912.

Trail End Construction Documents, Wyoming State Museum, Cheyenne.

Morell and Nichols. Letter to John B. Kendrick, 1 July 1911. Trail End Construction Documents, Wyoming State Museum, Cheyenne.

Morgan, Mary Kendrick. Letter to Manville Kendrick, 17 May 1960. Correspondence Files, Manville Kendrick Collection, Trail End State Historic Site, Sheridan, Wyoming.

Waid, D. Everett. Letter to Eula Wulfjen Kendrick, 25 May 1912. Trail End Construction Documents, Wyoming State Museum, Cheyenne.

Warren, Francis E. Letter to Chief of the Division of Military Aeronautics, 4 October 1918. Correspondence Files, Manville Kendrick Collection, Trail End State Historic Site, Sheridan, Wyoming.

_____. Letter to Manville Kendrick, 4 October 1918. Correspondence Files, Manville Kendrick Collection, Trail End State Historic Site, Sheridan, Wyoming.

❈ Index ❈

❋ About the Author ❋

Although born in Eureka, California, Cynde A. Georgen's roots are in northern Wyoming, her family having lived there since the early 1890s. After attending Northwest College in Powell and Montana State University in Bozeman, she obtained a bachelor's degree in history from the University of Wyoming in 1978.

Following a ten-year stint in broadcasting and publishing, during which she was named Associated Press Wyoming Broadcaster of the Year in 1988, Ms. Georgen began work at the Trail End State Historic Site. She was hired as curator in 1988 and has served as site superintendent since 1995. She and her staff received a National Award of Merit from the American Association for State and Local History in 1998 for their exhibit "Keeping the Home Fires Burning: Life on the Home Front During the Great War, 1917–1918."

Ms. Georgen has served as chair of the Colorado-Wyoming Association of Museums and as a member of the Sheridan County Historic Preservation Commission and the Sheridan Heritage Board. She and her husband, restoration carpenter Steve Baskin, make their home in Sheridan.